Aircraft Weight and Balance Handbook

FAA-H-8083-1A

U.S. DEPARTMENT OF TRANSPORTATION
Federal Aviation Administration
Flight Standards Service

Skyhorse Publishing

Skyhorse Publishing books may be purchased in bulk at special discounts for sales promotion, corporate gifts, fund-raising, or educational purposes. Special editions can also be created to specifications. For details, contact the Special Sales Department, Skyhorse Publishing, 555 Eighth Avenue, Suite 903, New York, NY 10018 or info@skyhorsepublishing.com.

www.skyhorsepublishing.com

10 9 8 7 6 5 4 3 2 1

Library of Congress Cataloging-in-Publication Data

Aircraft weight and balance handbook.
　　p. cm.
　"FAA-H-8083-1A, Federal Aviation Administration."
　Includes indexes.
　ISBN 978-1-61608-124-9 (pbk. : alk. paper)
　1. Stability of airplanes--Handbooks, manuals, etc. 2.
Airplanes--Weight--Handbooks, manuals, etc. I. United States.
Federal Aviation Administration.
　TL574.S7A37 2010
　629.134'52--dc22
　　　　　　　　　　　2010032762

Printed in China

Preface

FAA-H-8083-1A, *Aircraft Weight and Balance Handbook,* has been prepared in recognition of the importance of weight and balance technology in conducting safe and efficient flight. The objective of this handbook is twofold: to provide the Airframe and Powerplant Mechanic (A&P) with the method of determining the empty weight and empty-weight center of gravity (EWCG) of an aircraft, and to furnish the flightcrew with information on loading and operating the aircraft to ensure its weight is within the allowable limit and the center of gravity (CG) is within the allowable range.

Any time there is a conflict between the information in this handbook and specific information issued by an aircraft manufacturer, the manufacturer's data takes precedence over information in this handbook. Occasionally, the word *must* or similar language is used where the desired action is deemed critical. The use of such language is not intended to add to, interpret, or relieve a duty imposed by Title 14 of the Code of Federal Regulations (14 CFR).

This publication may be purchased from the Superintendent of Documents, U.S. Government Printing Office (GPO), Washington, DC 20402-9325, or from GPO's web site.

http://bookstore.gpo.gov

This handbook is also available for download, in pdf format, from the Regulatory Support Division's (AFS-600) web site.

http://www.faa.gov/about/office_org/headquarters_offices/ avs/offices/afs/afs600

This handbook is published by the U.S. Department of Transporation, Federal Aviation Administration, Airmen Testing Standards Branch, AFS-630, P.O. Box 25082, Oklahoma City, OK 73125.

Comments regarding this publication should be sent, in email form, to the following address.

AFS630Comments@faa.gov

Introduction

This handbook begins with the basic principle of aircraft weight and balance control, emphasizing its importance and including examples of documentation furnished by the aircraft manufacturer and by the FAA to ensure the aircraft weight and balance records contain the proper data.

Procedures for the preparation and the actual weighing of an aircraft are described, as are the methods of determining the location of the empty-weight center of gravity (EWCG) relative to both the datum and the mean aerodynamic chord (MAC).

Loading computations for general aviation aircraft are discussed, using both loading graphs and tables of weight and moment indexes.

Information is included that allows an A&P mechanic or repairman to determine the weight and center of gravity (CG) changes caused by repairs and alterations. This includes instructions for conducting adverse-loaded CG checks, also explaining the way to determine the amount and location of ballast needed to bring the CG within allowable limits.

The unique requirements for helicopter weight and balance control are discussed, including the determination of lateral CG and the way both lateral and longitudinal CG change as fuel is consumed.

A chapter is included giving the methods and examples of solving weight and balance problems, using hand-held electronic calculators, E6-B flight computers, and a dedicated electronic flight computer.

Contents

Weight and Balance Control

There are many factors that lead to efficient and safe operation of aircraft. Among these vital factors is proper weight and balance control. The weight and balance system commonly employed among aircraft consists of three equally important elements: the weighing of the aircraft, the maintaining of the weight and balance records, and the proper loading of the aircraft. An inaccuracy in any one of these elements nullifies the purpose of the whole system. The final loading calculations will be meaningless if either the aircraft has been improperly weighed or the records contain an error.

Improper loading cuts down the efficiency of an aircraft from the standpoint of altitude, maneuverability, rate of climb, and speed. It may even be the cause of failure to complete the flight, or for that matter, failure to start the flight. Because of abnormal stresses placed upon the structure of an improperly loaded aircraft, or because of changed flying characteristics of the aircraft, loss of life and destruction of valuable equipment may result.

The responsibility for proper weight and balance control begins with the engineers and designers, and extends to the aircraft mechanics that maintain the aircraft and the pilots who operate them.

Modern aircraft are engineered utilizing state-of-the-art technology and materials to achieve maximum reliability and performance for the intended category. As much care and expertise must be exercised in operating and maintaining these efficient aircraft as was taken in their design and manufacturing.

The designers of an aircraft have set the maximum weight, based on the amount of lift the wings or rotors can provide under the operation conditions for which the aircraft is designed. The structural strength of the aircraft also limits the maximum weight the aircraft can safely carry. The ideal location of the center of gravity (CG) was very carefully determined by the designers, and the maximum deviation allowed from this specific location has been calculated.

The manufacturer provides the aircraft operator with the empty weight of the aircraft and the location of its empty-weight center of gravity (EWCG) at the time the certified aircraft leaves the factory. Amateur-built aircraft must have this information determined and available at the time of certification.

The airframe and powerplant (A&P) mechanic or repairman who maintains the aircraft keeps the weight and balance records current, recording any changes that have been made because of repairs or alterations.

The pilot in command of the aircraft has the responsibility on every flight to know the maximum allowable weight of the aircraft and its CG limits. This allows the pilot to determine on the preflight inspection that the aircraft is loaded in such a way that the CG is within the allowable limits.

Weight Control

Weight is a major factor in airplane construction and operation, and it demands respect from all pilots and particular diligence by all A&P mechanics and repairmen. **Excessive weight reduces the efficiency of an aircraft and the safety margin available if an emergency condition should arise.**

When an aircraft is designed, it is made as light as the required structural strength will allow, and the wings or rotors are designed to support the maximum allowable weight. When the weight of an aircraft is increased, the wings or rotors must produce additional lift and the structure must support not only the additional static loads, but also the dynamic loads imposed by flight maneuvers. For example, the wings of a 3,000-pound airplane must support 3,000 pounds in level flight, but when the airplane is turned smoothly and sharply using a bank angle of 60°, the dynamic load requires the wings to support twice this, or 6,000 pounds.

Severe uncoordinated maneuvers or flight into turbulence can impose dynamic loads on the structure great enough to cause failure. In accordance with Title 14 of the Code of Federal Regulations (14 CFR) part 23, the structure of a normal category airplane must be strong enough to sustain a load factor of 3.8 times its weight. That is, every pound of weight added to an aircraft requires that the structure be strong enough to support an additional 3.8 pounds. An aircraft operated in the utility category must sustain a load factor of 4.4, and acrobatic category aircraft must be strong enough to withstand 6.0 times their weight.

The lift produced by a wing is determined by its airfoil shape, angle of attack, speed through the air, and the air density. When an aircraft takes off from an airport with a high density altitude, it must accelerate to a speed faster than would be required at sea level to produce enough lift to allow takeoff; therefore, a longer takeoff run is necessary. The distance needed may be longer than the available runway. When operating from a high-density altitude airport, the Pilot's Operating Handbook (POH) or Airplane Flight Manual (AFM) must be consulted to determine the maximum weight allowed for the aircraft under the conditions of altitude, temperature, wind, and runway conditions.

Effects of Weight

Most modern aircraft are so designed that if all seats are occupied, all baggage allowed by the baggage compartment is carried, and all of the fuel tanks are full, the aircraft will be grossly overloaded. This type of design requires the pilot to give great consideration to the requirements of the trip. If maximum range is required, occupants or baggage must be left behind, or if the maximum load must be carried, the range, dictated by the amount of fuel on board, must be reduced.

Some of the problems caused by overloading an aircraft are:

- the aircraft will need a higher takeoff speed, which results in a longer takeoff run.

- both the rate and angle of climb will be reduced.

- the service ceiling will be lowered.

- the cruising speed will be reduced.

- the cruising range will be shortened.

- maneuverability will be decreased.

- a longer landing roll will be required because the landing speed will be higher.

- excessive loads will be imposed on the structure, especially the landing gear.

The POH or AFM includes tables or charts that give the pilot an indication of the performance expected for any weight. An important part of careful preflight planning includes a check of these charts to determine the aircraft is loaded so the proposed flight can be safely made.

Weight Changes

The maximum allowable weight for an aircraft is determined by design considerations. However, the maximum operational weight may be less than the maximum allowable weight due to such considerations as high-density altitude or high-drag field conditions caused by wet grass or water on the runway. The maximum operational weight may also be limited by the departure or arrival airport's runway length.

One important preflight consideration is the distribution of the load in the aircraft. Loading the aircraft so the gross weight is less than the maximum allowable is not enough. This weight must be distributed to keep the CG within the limits specified in the POH or AFM.

If the CG is too far forward, a heavy passenger can be moved to one of the rear seats or baggage can be shifted from a forward baggage compartment to a rear compartment. If the CG is too far aft, passenger weight or baggage can be shifted forward. The fuel load should be balanced laterally: the pilot should pay special attention to the POH or AFM regarding the operation of the fuel system, in order to keep the aircraft balanced in flight.

Weight and balance of a helicopter is far more critical than for an airplane. With some helicopters, they may be properly loaded for takeoff, but near the end of a long flight when the fuel tanks are almost empty, the CG may have shifted enough for the helicopter to be out of balance laterally or longitudinally. Before making any long flight, the CG with the fuel available for landing must be checked to ensure it will be within the allowable range.

Airplanes with tandem seating normally have a limitation requiring solo flight to be made from the front seat in some airplanes or the rear seat in others. Some of the smaller helicopters also require solo flight be made from a specific seat, either the right, left, or center. These seating limitations will be noted by a placard, usually on the instrument panel, and they should be strictly adhered to.

As an aircraft ages, its weight usually increases due to trash and dirt collecting in hard-to-reach locations, and moisture absorbed in the cabin insulation. This growth in weight is normally small, but it can only be determined by accurately weighing the aircraft.

Changes of fixed equipment may have a major effect upon the weight of the aircraft. Many aircraft are overloaded by the installation of extra radios or instruments. Fortunately, the replacement of older, heavy electronic equipment with newer, lighter types results in a weight reduction. This weight change, however helpful, will probably cause the CG to shift and this must be computed and annotated in the weight and balance record.

Repairs and alteration are the major sources of weight changes, and it is the responsibility of the A&P mechanic or repairman making any repair or alteration to know the weight and location of these changes, and to compute the CG and record the new empty weight and EWCG in the aircraft weight and balance record.

If the newly calculated EWCG should happen to fall outside the EWCG range, it will be necessary to perform adverse loading check. This will require a forward and rearward adverse-loading check, and a maximum weight check. These weight and balance extreme conditions represent the maximum forward and rearward CG position for the aircraft. Adverse loading checks are a deliberate attempt to load an aircraft in a manner that will create the most critical balance condition and still remain within the design CG limits of the aircraft. If any of the checks fall outside the loaded CG range, the aircraft must be reconfigured or placarded to prevent the pilot from loading the aircraft improperly. It is sometimes possible to install fixed ballast in order for the aircraft to again operate within the normal CG range.

The A&P mechanic or repairman conducting an annual or condition inspection must ensure the weight and balance data in the aircraft records is current and accurate. It is the responsibility of the pilot in command to use the most current weight and balance data when operating the aircraft.

Stability and Balance Control

Balance control refers to the location of the CG of an aircraft. This is of primary importance to aircraft stability, which determines safety in flight.

The CG is the point at which the total weight of the aircraft is assumed to be concentrated, and the CG must be located within specific limits for safe flight. Both lateral and longitudinal balance are important, but the prime concern is longitudinal balance; that is, the location of the CG along the longitudinal or lengthwise axis.

An airplane is designed to have stability that allows it to be trimmed so it will maintain straight and level flight with hands off the controls. Longitudinal stability is maintained by ensuring the CG is slightly ahead of the center of lift. This produces a fixed nose-down force independent of the airspeed. This is balanced by a variable nose-up force, which is produced by a downward aerodynamic force on the horizontal tail surfaces that varies directly with the airspeed. [Figure 1-1]

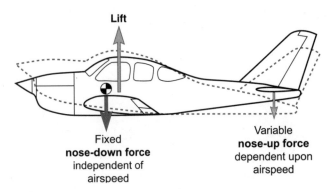

Figure 1-1. *Longitudinal forces acting on an airplane in flight.*

If a rising air current should cause the nose to pitch up, the airplane will slow down and the downward force on the tail will decrease. The weight concentrated at the CG will pull the nose back down. If the nose should drop in flight, the airspeed will increase and the increased downward tail load will bring the nose back up to level flight.

As long as the CG is maintained within the allowable limits for its weight, the airplane will have adequate longitudinal stability and control. If the CG is too far aft, it will be too near the center of lift and the airplane will be unstable, and difficult to recover from a stall. [Figure 1-2] If the unstable airplane should ever enter a spin, the spin could become flat and recovery would be difficult or impossible.

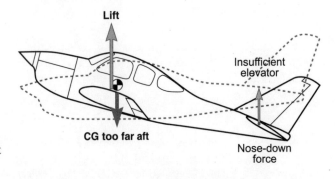

Figure 1-2. *If the CG is too far aft at the low stall airspeed, there might not be enough elevator nose-down authority to get the nose down for recovery.*

If the CG is too far forward, the downward tail load will have to be increased to maintain level flight. This increased tail load has the same effect as carrying additional weight;

the aircraft will have to fly at a higher angle of attack, and drag will increase.

A more serious problem caused by the CG being too far forward is the lack of sufficient elevator authority. At slow takeoff speeds, the elevator might not produce enough nose-up force to rotate and on landing there may not be enough elevator force to flare the airplane. [Figure 1-3] Both takeoff and landing runs will be lengthened if the CG is too far forward.

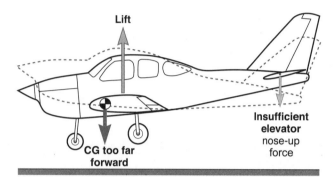

Figure 1-3. *If the CG is too far forward, there will not be enough elevator nose-up force to flare the airplane for landing.*

The basic aircraft design assumes that lateral symmetry exists. For each item of weight added to the left of the centerline of the aircraft (also known as buttock line zero, or BL-0), there is generally an equal weight at a corresponding location on the right.

The lateral balance can be upset by uneven fuel loading or burnoff. The position of the lateral CG is not normally computed for an airplane, but the pilot must be aware of the adverse effects that will result from a laterally unbalanced condition. [Figure 1-4] This is corrected by using the aileron trim tab until enough fuel has been used from the tank on the heavy side to balance the airplane. The deflected trim tab deflects the aileron to produce additional lift on the heavy side, but it also produces additional drag, and the airplane flies inefficiently.

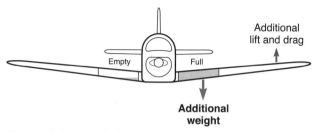

Figure 1-4. *Lateral imbalance causes wing heaviness, which may be corrected by deflecting the aileron. The additional lift causes additional drag and the airplane flies inefficiently.*

Helicopters are affected by lateral imbalance more than airplanes. If a helicopter is loaded with heavy occupants and fuel on the same side, it could be out of balance enough to make it unsafe to fly. It is also possible that if external loads are carried in such a position to require large lateral displacement of the cyclic control to maintain level flight, the fore-and-aft cyclic control effectiveness will be limited.

Sweptwing airplanes are more critical due to fuel imbalance because as the fuel is used from the outboard tanks, the CG shifts forward, and as it is used from the inboard tanks, the CG shifts aft. [Figure 1-5] For this reason, fuel-use scheduling in sweptwing airplanes operation is critical.

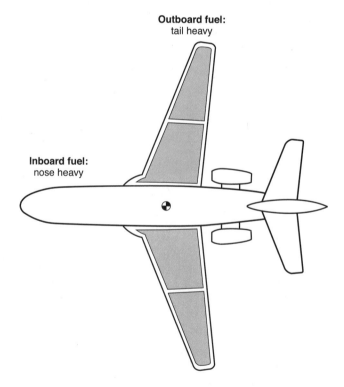

Figure 1-5. *Fuel in the tanks of a sweptwing airplane affects both lateral and longitudinal balance. As fuel is used from an outboard tank, the CG shifts forward.*

Weight Control for Aircraft other than Fixed and Rotorwing

Some light aircraft utilize different methods of determining weight and balance from the traditional fixed and rotorwing aircraft. These aircraft achieve flight control differently than the fixed-wing airplane or helicopter. Most notable of these are weight shift control (WSC) aircraft (also known as trikes), powered parachutes, and balloons.

These aircraft typically do not specify either an empty weight center of gravity or a center of gravity range. They require only a certified or approved maximum weight.

To understand why this is so, a look at how flight control is achieved is helpful.

As an example, airplanes and WSC aircraft both control flight under the influence of the same four forces (lift, gravity, thrust, and drag), and around the same three axes (pitch, yaw, and roll). However, each aircraft accomplishes this control in a very different manner. This difference helps explain why the fixed-wing airplane requires an established weight and a known center of gravity, whereas the WSC aircraft only requires the known weight.

The fixed-wing airplane has moveable controls that alter the lift on various airfoil surfaces to vary pitch, roll, and yaw. These changes in lift, in turn, change the characteristics of the flight parameters. Weight normally decreases in flight due to fuel consumption, and the airplane center of gravity changes with this weight reduction. An airplane utilizes its variable flight controls to compensate and maintain controllability through the various flight modes and as the center of gravity changes. An airplane has a center of gravity range or envelope within which it must remain if the flight controls are to remain effective and the airplane safely operated.

The WSC aircraft has a relatively set platform wing without a tail. The pilot, achieves control by shifting weight. In the design of this aircraft, the weight of the airframe and its payload is attached to the wing at a single point in a pendulous arrangement. The pilot through the flight controls, controls the arm of this pendulum and thereby controls the aircraft. When a change in flight parameter is desired, the pilot displaces the aircraft's weight in the appropriate distance and direction. This change momentarily disrupts the equilibrium between the four forces acting on the aircraft. The wing, due to its inherent stability, then moves appropriately to re-establish the desired relationship between these forces. This happens by the wing flexing and altering its shape. As the shape is changed, lift is varied at different points on the wing to achieve the desired flight parameters.

The flight controls primarily affect the pitch-and-roll axis. Since there is no vertical tail plane, minimal or no ability exists to directly control yaw. However, unlike the airplane, the center of gravity experienced by the wing remains constant. Since the weight of the airframe acts through the single point (wing attach point), the range over which the weight may act is fixed at the pendulum arm or length. Even though the weight decreases as fuel is consumed, the weight remains focused at the wing attach point. Most importantly, because the range is fixed, the need to establish a calculated range is not required.

The powered parachute also belongs to the pendulum-style aircraft. Its airframe center of gravity is fixed at the pendulum attach point. It is more limited in controllability than the WSC aircraft because it lacks an aerodynamic pitch control. Pitch (and lift) control is primarily a function of the power control. Increased power results in increased lift; cruise power amounts to level flight; decreased power causes a descent. Due to this characteristic, the aircraft is basically a one-air speed aircraft. Once again, because the center of gravity is fixed at the attach point to the wing, there can be no center of gravity range.

Roll control on a powered parachute is achieved by changing the shape of the wing. The change is achieved by varying the length of steering lines attached to the outboard trailing edges of the wing. The trailing edge of the parachute is pulled down slightly on one side or the other to create increased drag along that side. This change in drag creates roll and yaw, permitting the aircraft to be steered.

The balloon is controlled by the pilot only in the vertical dimension; this is in contrast to all other aircraft. He or she achieves this control through the use of lift and weight. Wind provides all other movement. The center of gravity of the gondola remains constant beneath the balloon envelope. As in WSC and powered-parachute aircraft, there is no center of gravity limitation.

Aircraft can perform safely and achieve their designed efficiency only when they are operated and maintained in the way their designers intended. This safety and efficiency is determined to a large degree by holding the aircraft's weight and balance parameters within the limits specified for its design. The remainder of this handbook describes the way in which this is done.

Weight and Balance

Chapter 2

Theory and Documentation

Weight and Balance Theory

Two elements are vital in the weight and balance considerations of an aircraft.

- The total weight of the aircraft must be no greater than the maximum weight allowed by the FAA for the particular make and model of the aircraft.

- The center of gravity, or the point at which all of the weight of the aircraft is considered to be concentrated, must be maintained within the allowable range for the operational weight of the aircraft.

Aircraft Arms, Weight, and Moments

The term arm, usually measured in inches, refers to the distance between the center of gravity of an item or object and the datum. Arms ahead of, or to the left of the datum are negative(-), and those behind, or to the right of the datum are positive(+). When the datum is ahead of the aircraft, all of the arms are positive and computational errors are minimized. Weight is normally measured in pounds. When weight is removed from an aircraft, it is negative(-), and when added, it is positive (+).

The manufacturer establishes the maximum weight and range allowed for the CG, as measured in inches from the reference plane called the datum. Some manufacturers specify this range as measured in percentage of the mean aerodynamic chord (MAC), the leading edge of which is located a specified distance from the datum.

The datum may be located anywhere the manufacturer chooses; it is often the leading edge of the wing or some specific distance from an easily identified location. One popular location for the datum is a specified distance forward of the aircraft, measured in inches from some point, such as the nose of the aircraft, or the leading edge of the wing, or the engine firewall.

The datum of some helicopters is the center of the rotor mast, but this location causes some arms to be positive and others negative. To simplify weight and balance computations, most modern helicopters, like airplanes,

have the datum located at the nose of the aircraft or a specified distance ahead of it.

A moment is a force that tries to cause rotation, and is the product of the arm, in inches, and the weight, in pounds. Moments are generally expressed in pound-inches (lb-in) and may be either positive or negative. Figure 2-1 shows the way the algebraic sign of a moment is derived. Positive moments cause an airplane to nose up, while negative moments cause it to nose down.

Weight	Arm	Moment	Rotation
+	+	+	Nose up
+	−	−	Nose down
−	+	−	Nose down
−	−	+	Nose up

Figure 2-1. *Relationships between the algebraic signs of weight, arms, and moments.*

The Law of the Lever

The weight and balance problems are based on the physical law of the lever. This law states that a lever is balanced when the weight on one side of the fulcrum multiplied by its arm is equal to the weight on the opposite side multiplied by its arm. In other words, the lever is balanced when the algebraic sum of the moments about the fulcrum is zero. [Figure 2-2] This is the condition in which the positive moments (those that try to rotate the lever clockwise) are equal to the negative moments (those that try to rotate it counter-clockwise).

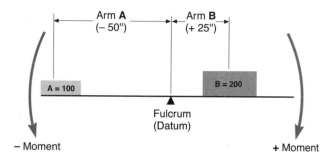

Figure 2-2. *The lever is balanced when the algebraic sum of the moments is zero.*

Consider these facts about the lever in Figure 2-2: The 100-pound weight A is located 50 inches to the left of the fulcrum (the datum, in this instance), and it has a moment of 100 X -50 = -5,000 in-lb. The 200-pound weight B is located 25 inches to the right of the fulcrum, and its moment is 200 x +25 = +5000 in-lb. The sum of the moment is -5000 + 5000 = 0, and the lever is balanced. [Figure 2-3] The forces that try to rotate it clockwise have the same magnitude as those that try to rotate it counterclockwise.

Item	Weight (lb)	Arm (in)	Moment (lb-in)
Weight A	100	−50	−5,000
Weight B	200	+25	+5,000
	300		0

Figure 2-3. *When a lever is in balance, the sum of the moments is zero.*

Determining the CG

One of the easiest ways to understand weight and balance is to consider a board with weights placed at various locations. We can determine the CG of the board and observe the way the CG changes as the weights are moved.

The CG of a board like the one in Figure 2-4 may be determined by using these four steps:

1. Measure the arm of each weight in inches from the datum.

2. Multiply each arm by its weight in pounds to determine the moment in pound-inches of each weight.

3. Determine the total of all weights and of all the moments. Disregard the weight of the board.

4. Divide the total moment by the total weight to determine the CG in inches from the datum.

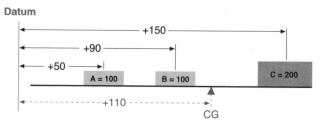

Figure 2-4. *Determining the center of gravity from a datum located off the board.*

In Figure 2-4, the board has three weights, and the datum is located 50 inches to the left of the CG of weight A. Determine the CG by making a chart like the one in Figure 2-5.

Item	Weight	Arm	Moment	CG
Weight A	100	50	5,000	
Weight B	100	90	9,000	
Weight C	200	150	30,000	
	400		44,000	110

Figure 2-5. *Determining the CG of a board with three weights and the datum located off the board.*

As noted in Figure 2-5, A weighs 100 pounds and is 50 inches from the datum: B weighs 100 pounds and is 90 inches from the datum; C weighs 200 pounds and is 150 inches from the datum. Thus the total of the three weights is 400 pounds, and the total moment is 44,000 lb-in.

Determine the CG by dividing the total moment by the total weight.

$$CG = \frac{\text{Total moment}}{\text{Total weight}}$$

$$= \frac{44,000}{400}$$

$$= 110 \text{ inches from the datum}$$

To prove this is the correct CG, move the datum to a location 110 to the right of the original datum and determine the arm of each weight from this new datum, as in Figure 2-6. Then make a new chart similar to the one in Figure 2-7. If the CG is correct, the sum of the moments will be zero.

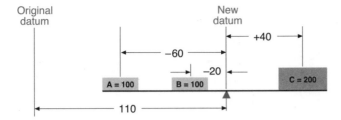

Figure 2-6. *Arms from the datum assigned to the CG.*

The new arm of weight A is 110 - 50 = 60 inches, and since this weight is to the left of the datum, its arm is negative, or -60 inches. The new arm of weight B is 110 - 90 = 20 inches, and it is also to the left of the datum, so it is - 20; the new arm of weight C is 150 - 110 = 40 inches. It is to the right of the datum and is therefore positive.

Item	Weight	Arm	Moment
Weight A	100	−60	−6,000
Weight B	100	−20	−2,000
Weight C	200	+40	+8,000
			0

Figure 2-7. *The board balances at a point 110 inches to the right of the original datum.*

The board is balanced when the sum of the moments is zero. The location of the datum used for determining the arms of the weights is not important; it can be anywhere. But all of the measurements must be made from the same datum location.

Determining the CG of an airplane is done in the same way as determining the CG of the board in the previous example. [Figure 2-8] Prepare the airplane for weighing (as explained in Chapter 3) and place it on three scales. All tare weight, that is, the weight of any chocks or devices used to hold the aircraft on the scales, is subtracted from the scale reading, and the net weight from each wheel weigh point is entered on the chart like the one in Figure 2-9. The arms of the weighing points are specified in the Type Certificate Data Sheet (TCDS) for the airplane in terms of stations, which are distances in inches from the datum. Tare weight also includes items used to level the aircraft.

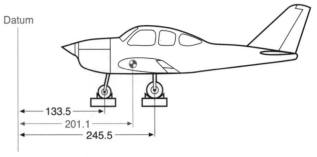

Figure 2-8. *Determining the CG of an airplane whose datum is ahead of the airplane.*

Item	Weight	Arm	Moment	CG
Main wheels	3,540	245.5	869,070	
Nose wheel	2,322	133.5	309,987	
Total	5,862		1,179,057	201.1

Figure 2-9. *Chart for determining the CG of an airplane whose datum is ahead of the airplane.*

The empty weight of this aircraft is 5,862 pounds. Its EWCG, determined by dividing the total moment by the total weight, is located at fuselage station 201.1. This is 201.1 inches behind the datum.

$$CG = \frac{\text{Total moment}}{\text{Total weight}}$$

$$= \frac{1,179,057}{5,862}$$

$$= 201.1 \text{ inches behind the datum}$$

Shifting the CG

One common weight and balance problem involves moving passengers from one seat to another or shifting baggage or cargo from one compartment to another to move the CG to a desired location. This also can be visualized by using a board with three weights and then working out the problem the way it is actually done on an airplane.

Solution by Chart

The CG of a board can be moved by shifting the weights as demonstrated in Figure 2-10. As the board is loaded, it balances at a point 72 inches from the CG of weight A. [Figure 2-11]

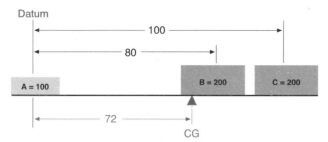

Figure 2-10. *Moving the CG of a board by shifting the weights. This is the original configuration.*

Item	Weight	Arm	Moment	CG
Weight A	100	0	0	
Weight B	200	80	16,000	
Weight C	200	100	20,000	
	500		36,000	72

Figure 2-11. *Shifting the CG of a board by moving one of the weights. This is the original condition of the board.*

To shift weight B so the board will balance about its center, 50 inches from the CG of weight A, first determine the arm of weight B that will produce a moment that causes the total moment of all three weights around this desired balance point to be zero. The combined moment of weights A and C around this new balance point, is 5,000 in-lb, so the moment of weight B will have to be -5,000 lb-in in order for the board to balance. [Figure 2-12]

Item	Weight	Arm	Moment
Weight A	100	−50	−5,000
Weight B			
Weight C	200	+50	+10,000
			+5,000

Figure 2-12. *Determining the combined moment of weights A and C.*

Determine the arm of weight B by dividing its moment, -5,000 lb-in, by its weight of 200 pounds. Its arm is -25 inches.

$$\text{Arm B} = \frac{\text{Moment}}{\text{Weight}}$$

$$= \frac{-5,000}{200}$$

$$= -25$$

To balance the board at its center, weight B will have to be placed so its CG is 25 inches to the left of the center of the board, as in Figure 2-13.

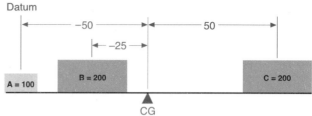

Figure 2-13. *Placement of weight B to cause the board to balance about its center.*

Basic Weight and Balance Equation

$$\frac{\text{Weight to be shifted}}{\text{Total weight}} = \frac{\Delta CG}{\text{Distance weight is shifted}}$$

This equation can be rearranged to find the distance a weight must be shifted to give a desired change in the CG location:

$$\text{Distance weight is shifted} = \frac{\text{Total weight} \times \Delta CG}{\text{Weight shifted}}$$

This equation can also be rearranged to find the amount of weight to shift to move the CG to a desired location:

$$\text{Weight shifted} = \frac{\text{Total weight} \times \Delta CG}{\text{Distance weight is shifted}}$$

It can also be rearranged to find the amount the CG is moved when a given amount of weight is shifted:

$$\Delta CG = \frac{\text{Weight shifted} \times \text{Distance weight is shifted}}{\text{Total weight}}$$

Finally, this equation can be rearranged to find the total weight that would allow shifting a given amount of weight to move the CG a given distance:

$$\text{Total weight} = \frac{\text{Weight shifted} \times \text{Distance weight is shifted}}{\Delta CG}$$

Solution by Formula

This same problem can also be solved by using this basic equation:

$$\frac{\text{Weight to be shifted}}{\text{Total weight}} = \frac{\Delta CG}{\text{Distance weight is shifted}}$$

Rearrange this formula to determine the distance weight B must be shifted:

$$\text{Distance weight B is shifted} = \frac{\text{Total weight} \times \Delta CG}{\text{Weight shifted}}$$

$$= \frac{500 \times -22}{200}$$

$$= -55 \text{ inches}$$

The CG of the board in Figure 2-10 was 72 inches from the datum. This CG can be shifted to the center of the board as in Figure 2-13 by moving weight B. If the 200-pound weight B is moved 55 inches to the left, the CG will shift from 72 inches to 50 inches, a distance of 22 inches. The sum of the moments about the new CG will be zero. [Figure 2-14]

Item	Weight	Arm	Moment
Weight A	100	−50	−5,000
Weight B	200	−25	−5,000
Weight C	200	+50	+10,000
			0

Figure 2-14. *Proof that the board balances at its center. The board is balanced when the sum of the moments is zero.*

When the distance the weight is to be shifted is known, the amount of weight to be shifted to move the CG to any location can be determined by another arrangement of the basic equation. Use the following arrangement of the formula to determine the amount of weight that will have to be shifted from station 80 to station 25, to move the CG from station 72 to station 50.

$$\text{Weight shifted} = \frac{\text{Total weight} \times \Delta CG}{\text{Distance weight is shifted}}$$

$$= \frac{500 \times 22}{55}$$

$$= 200 \text{ pounds}$$

If the 200-pound weight B is shifted from station 80 to station 25, the CG will move from station 72 to station 50.

A third arrangement of this basic equation may be used to determine the amount the CG is shifted when a given amount of weight is moved for a specified distance (as it was done in Figure 2-10). Use this formula to determine the amount the CG will be shifted when 200-pound weight B is moved from +80 to +25.

$$\Delta CG = \frac{\text{Weight shifted} \times \text{Distance it is shifted}}{\text{Total weight}}$$

$$= \frac{200 \times 55}{500}$$

$$= 22 \text{ inches}$$

Moving weight B from +80 to +25 will move the CG 22 inches, from its original location at +72 to its new location at +50 as seen in Figure 2-13.

Shifting the Airplane CG

The same procedures for shifting the CG by moving weights can be used to change the CG of an airplane by rearranging passengers or baggage.

Consider this airplane:

Airplane empty weight and EWCG 1340 lbs @ +37.0
Maximum gross weight .... 2,300 lbs
CG limits .... +35.6 to +43.2
Front seats (2) .... +35
Rear seats (2) .... +72
Fuel .... 40 gal @ +48
Baggage (maximum) .... 60 lbs @ +92

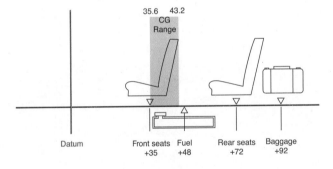

Figure 2-15. *Loading diagram for a typical single-engine airplane.*

The pilot has prepared a chart, Figure 2-16, with certain permanent data filled in and blanks left to be filled in with information on this particular flight.

Item	Weight 2,300 max	Arm	Moment	CG +35.6 to +43.2
Airplane	1,340	37	49,580	
Front Seats		35		
Rear Seats		72		
Fuel		48		
Baggage		92		

Figure 2-16. *Blank loading chart.*

For this flight, the 140-pound pilot and a 115-pound passenger are to occupy the front seats, and a 212-pound and a 97-pound passenger are in the rear seats. There will be 50 pounds of baggage, and the flight is to have maximum range, so maximum fuel is carried. The loading chart, Figure 2-17, is filled in using the information from Figure 2-15.

Item	Weight 2,300 max	Arm	Moment	CG +35.6 to +43.2
Airplane	1,340	37	49,580	
Front Seats	255	35	8,925	
Rear Seats	309	72	22,248	
Fuel	240	48	11,520	
Baggage	50	92	4,600	
	2,194		96,873	44.1

Figure 2-17. *This completed loading chart shows the weight is within limits, but the CG is too far aft.*

With this loading, the total weight is less than the maximum of 2,300 pounds and is within limits, but the CG is 0.9 inch too far aft.

One possible solution would be to trade places between the 212-pound rear-seat passenger and the 115-pound front-seat passenger. Use a modification of the basic weight and balance equation to determine the amount the CG will change when the passengers swap seats.

$$\Delta CG = \frac{\text{Weight shifted} \times \text{Distance it is shifted}}{\text{Total weight}}$$

$$= \frac{(212 - 115) \times (72 - 35)}{2,194}$$

$$= \frac{97 \times 37}{2,194}$$

$$= 1.6 \text{ inches}$$

The two passengers changing seats moved the CG forward 1.6 inches, which places it within the operating range. This can be proven correct by making a new chart incorporating the changes. [Figure 2-18]

Item	Weight 2,300 max	Arm	Moment	CG +35.6 to +43.2
Airplane	1,340	37	49,580	
Front Seats	352	35	12,320	
Rear Seats	212	72	15,264	
Fuel	240	48	11,520	
Baggage	50	92	4,600	
	2,194		93,284	42.5

Figure 2-18. *This loading chart, made after the seat changes, shows both the weight and balance are within allowable limits.*

Weight and Balance Documentation

FAA-Furnished Information

Before an aircraft can be properly weighed and its empty-weight center of gravity computed, certain information must be known. This information is furnished by the FAA to anyone for every certificated aircraft in the Type Certificate Data Sheets (TCDS) or Aircraft Specifications and can be accessed via the internet at: www.faa.gov (home page), from that page, select "Regulations and Policies," and at that page, select "Regulatory and Guidance Library." This is the official FAA technical reference library.

When the design of an aircraft is approved by the FAA, an Approved Type Certificate and TCDS are issued. The TCDS includes all of the pertinent specifications for the aircraft, and at each annual or 100-hour inspection, it is the responsibility of the inspecting mechanic or repairman to ensure that the aircraft adheres to them. See pages 2-7 through 2-9, for examples of TCDS excerpts. A note about the TCDS: aircraft certificated before January 1, 1958, were issued Aircraft Specifications under the Civil Air Regulations (CARs), but when the Civil Aeronautical Administration (CAA) was replaced by the FAA, Aircraft Specifications were replaced by the Type Certificate Data Sheets. The weight and balance information on a TCDS includes the following:

Data Pertinent to Individual Models

This type of information is determined in the sections pertinent to each individual model:

CG Range
Normal Category
(+82.0) to (+93.0) at 2,050 pounds.
(+87.4) to (+93.0) at 2,450 pounds.

Utility Category
(+82.0) to (+86.5) at 1,950 pounds.
Straight-line variations between points given.

DEPARTMENT OF TRANSPORTATION
FEDERAL AVIATION ADMINISTRATION

2A13	
Revision 41	
PIPER	
PA-28-140	PA-28-151
PA-28-150	PA-28-181
PA-28-160	PA-28-161
PA-28-180	PA-28R-201
PA-28-235	PA-28R-201T
PA-28S-160	PA-28-236
PA-28S-180	PA-28RT-201
PA-28R-180	PA-28RT-201T
PA-28R-200	PA-28-201T
May 12, 1987	

TYPE CERTIFICATE DATA SHEET NO. 2A13

This data sheet, which is a part of Type Certificate 2A13, prescribes conditions and limitations under which the product for which the type certificate was issued meets the airworthiness requirements of the Civil Air Regulations.

Type Certificate Holder   Piper Aircraft Corporation
2926 Piper Drive
Vero Beach, Florida 32960

I. Model PA-28-160, Cherokee, 4 PCLM (Normal Category), Approved October 31, 1960.

Engine
Lycoming 0-320-B2B or 0-320-D2a with Carburetor setting 10-3678-32

Fuel
91/96 minimum grade aviation gasoline.

Engine Limits
For all operations, 2700 r.p.m. (160 h.p.)

Propeller and Propeller Limits
Sensenich M74DM or 74DM6 on S/N 1 through 1760 1760A; Sensenich M74DMS or 74D6S5 on S/N 1761 and up. Static r.p.m. at maximum permission throttle setting: Not over 2425, not under 2325. No additional tolerance permitted.
Diameter: Not over 74", not under 72.5".
See Note 10.

Propeller Spinner
Piper P/N 14422-00 on S/N 1 through 1760A;
Piper P/N 63760-04 or 65805 on S/N 1761 and up.
See Note 11.

Page No.	1	2	3	4	5	6	7	8	9	10	11	12	13	14	15	16	17	18	19
Rev. No.	41	36	36	35	35	36	36	35	36	36	35	35	36	35	36	35	36	36	36
Page No.	20	21	22	23	24	25	26	27	28	29	30	31	32	33	34	35	36	37	38
Rev. No.	36	35	36	36	35	37	38	37	39	37	37	38	39	38	41	38	39	38	41
Page No.	39	40	41	42	43	44	45	46											
Rev. No.	38	41	38	38	38	38	38	39											

Figure 2-19. *Excerpts from a Type Certificate Data Sheet.*

Airspeed Limits	Never exceed	171 m.p.h. (148 knots) CAS
	Maximum Structural	140 m.p.h. (121 knots) CAS
	cruising	140 m.p.h. (121 knots) CAS
	Maneuvering	129 m.p.h. (112 knots) CAS
	Flaps extended	115 m.p.h. (100 knots) CAS

Center of Gravity Range
(+84.0) to (+95.9) at 1650 lb. or less
(+85.9) to (+95.9) at 1975 lb.
(+88.2) to (+95.9) at 2200 lb.
Straight line variation between points given

Empty Wt. C.G. Range None
Maximum Weight 2200 lb.
No. of Seats 4 (2 at +85.5, 2 at +118.1)
Maximum Baggage 125 lbs. (+142.8) (S/N 28-1 through 28-1760A)
See NOTE 8.
200 lbs. (+142.8) (S/N 28-1761 and up)

Fuel Capacity
50 gal. (2 wing tanks) (+95)
See NOTE 1 for data on system fuel.

Oil Capacity
8 qts. (+32.5), 6 qts. useable
See NOTE 1 for data on system oil.

Control Surface Movements	Wing flaps	(±2°)	Up 0°	Down	40°
	Ailerons	(±2°)	Up 30°	Down	15°
	Rudder	(±2°)	Left 27°	Right	27°
	Stabilator	(±2°)	Up 18°	Down	2°
	Stabilator tab	(±1°)	Up 3°	Down	12°

Nose Wheel Travel
(+1°) Left 30° Right 30°
(Effective on S/N 1 through 3377)
Left 22° Right 22°
(Effective on S/N 3378 and up)

Manufacturer's Serial Nos.
28-03, 28-1 and up.

II. Model PA-28-150, Cherokee, 4 PCLM (Normal Category), Approved June 2, 1961

Engine Lycoming 0-320-A2B or 0-320-E2A with carburetor setting 10-3678-32
Fuel 80/87 minimum grade aviation gasoline
Engine Limits For all operations, 2700 r.p.m. (150 h.p.)

Propeller and Propeller Limits
Sensenich M74DM or 74DM6 on S/N 1 through 1760A;
Sensenich M74DMS or 74DM6S5 on S/N 1761 and up
Static r.p.m. at maximum permissible throttle setting not over 2375, not under 2275.
No additional tolerance permitted.
Diameter: Not over 74", not under 72.5."
See NOTE 10.

Figure 2-19. *Excepts for a Type Certificate Data Sheet (continued)*

Data Pertinent to All Models:

Datum

78.4" forward of wing leading edge (straight wing only). 78.4" forward of inboard intersection of straight and tapered sections (semi-tapered wings).

Leveling Means

Two screws left side fuselage below window.

Certification Basis

Type Certificate No. 2A13 issued October 31, 1960. Date of Application for Type Certificate, February 14, 1965.

Delegation Option Authorization granted per FAR 21, Subpart J. July 17, 1968.

PA-28-140 and PA-28-151: CAR 3, effective May 15, 1956, including Amendments 3-1, 3-2, 3-4, and paragraphs 3.304 and 3.705 of Amendment 3-7.

PA-28-150, PA-28-160, PA-28-180, PA-28-235, PA-28S-160, PA-28S-180, PA-28R-180, PA-28R-200; CAR 3, effective May 15, 1956, including Amendments 3-1, 3-2 and paragraphs 3.304 and 3.705.

PA-28-161: CAR 3 effective May 15, 1956, through Amendment 3-2; paragraph 3.387(d) of Amendment 3-4; paragraphs 3.304 and 3.705 of Amendment 3-7; FAR 23.959 of Amendment 23-7; FAR 36 effective December 1, 1969, through Amendment 36-4.

PA-28-181: CAR 3 effective May 15, 1956, through Amendment 3-2, Amendment 3-4 and paragraphs 3.304 and 3.705 of Amendment 3-7. Also, FAR 23.207, 23.221 and 23.959 of Amendment 23-7.

PA-28R-201: CAR 3 effective May 15, 1956, through Amendment 3-2; paragraphs 3.304 and 3.705 of Amendment 3-7; paragraphs 23.221, 23.959, 23.965, 23.967(e)(2), 23.1091 and 23.1093 of FAR 23 Amendment 23-16; FAR 36 effective December 1, 1969, through Amendment 36-4 (no acoustical change).

PA-28R-201T: CAR 3 effective May 15, 1956, through Amendment 3-2 including paragraphs 3.304 and 3.705 of Amendment 3-7; FAR 23.221, 23.901, 23.909, 23.959, 23.965, 23.967(e)(2), 23.1041, 23.1043, 23.1047, 23.1143, 23.1305, 23.1441 and 23.1527 of Amendment 23-16; FAR 36 effective December 1, 1969, through Amendment 36-4.

PA-28-236: CAR 3 effective May 15, 1956, through Amendment 3-2, and paragraphs 3.304 and 3.705 of Amendment 3-7 effective May 3, 1962. FAR 23.221, 23.959, 23.1091, and 23.1093 of FAR Part 23, Amendment 23-17 effective February 1, 1977; FAR 23.1581(b)(2) of FAR 23 Amendment 23-21 effective March 1, 1978; and applicable portions of FAR 36, as amended up to Amendment 36-9 effective April 3, 1978.

Figure 2-19. *Excepts for a Type Certificate Data Sheet (continued)*

If this information is given, there may be a chart on the TCDS similar to the one in Figure 2-20. This chart helps visualize the CG range. Draw a line horizontally from the aircraft weight and a line vertically from the fuselage station on which the CG is located. If these lines cross inside the enclosed area, the CG is within the allowable range for the weight.

Note that there are two enclosed areas: the larger is the CG range when operating in the Normal category only, and the smaller range is for operating in both the Normal and Utility categories. When operating with the weight and CG limitations shown for Utility category, the aircraft is approved for limited acrobatics such as spins, lazy eights, chandelles, and steep turns in which the bank angle exceeds 60°. When operating outside of the smaller enclosure but within the larger, the aircraft is restricted from these maneuvers.

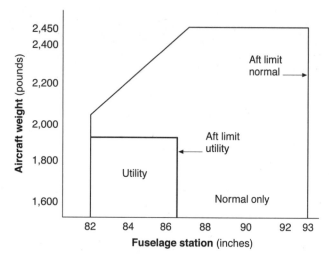

Figure 2-20. *CG range chart.*

If the aircraft has retractable landing gear, a note may be added, for example:

"Moment due to retracting of landing gear (+819 lb-in)."

Empty Weight CG Range

When all of the seats and baggage compartments are located close together, it is not possible, as long as the EWCG is located within the EWCG range, to legally load the aircraft so that its operational CG falls outside this allowable range. If the seats and baggage areas extend over a wide range, the EWCG range will be listed as "None."

Maximum Weights

The maximum allowable takeoff and landing weights and the maximum allowable ramp weight are given. This basic information may be altered by a note, such as the following:

"NOTE 5. A landing weight of 6,435 lbs must be observed if 10 PR tires are installed on aircraft not equipped with 60-810012-15 (LH) or 60-810012-16 (RH) shock struts."

Number of Seats

The number of seats and their arms are given in such terms as:

"4 (2 at +141, 2 at +173)"

Maximum Baggage (Structural Limit)

This is given as:

"500 lbs at +75 (nose compartment)
655 lbs at +212 (aft area of cabin)"

Fuel Capacity

This important information is given in such terms as:

"142 gal (+138) comprising two interconnected cells in each wing"

-or

"204 gal (+139) comprising three cells in each wing and one cell in each nacelle (four cells interconnected) See NOTE 1 for data on fuel system."

"NOTE 1" will read similar to the following example:

"NOTE 1. Current weight and balance data, including list of equipment included in standard empty weight and loading instructions when necessary, must be provided for each aircraft at the time of original certification.

The standard empty weight and corresponding center of gravity locations must include unusable fuel of 24 lbs at (+135)."

Oil Capacity (Wet Sump)

The quantity of the full oil supply and its arm are given in such terms as:

"26 qt (+88)"

Data Pertinent to all Models

Datum

The location of the datum may be described, for example, as:

"Front face of firewall"

-or

78.4 inches forward of wing leading edge (straight wing only).
78.4 inches forward of inboard intersection of straight and tapered sections (semi-tapered wings).

Leveling Means

A typical method is:

"Upper door sill."

This means that a spirit level is held against the upper door sill and the aircraft is level when the bubble is centered. Other methods require a spirit level to be placed across leveling screws or leveling lugs in the primary aircraft structure or dropping a plumbline between specified leveling points.

TCDS are issued for aircraft that have been certificated since January 1, 1958, when the FAA came into being. For aircraft certificated before this date, basically the same data is included in Aircraft, Engine, or Propeller Specifications that were issued by the Civil Aeronautics Administration.

Within the Type Certificate Data Sheets, Specifications, and Listings, Volume VI, titled "The Aircraft Listings" includes weight and balance information on aircraft of which there are fewer than 50 listed as being certificated.

Manufacturer-Furnished Information

When an aircraft is initially certificated, its empty weight and EWCG are determined and recorded in the weight and balance record such as the one in Figure 2-21. Notice in this figure that the moment is expressed as "Moment (lb-in/1000)." This is a moment index which means that the moment, a very large number, has been divided by 1,000 to make it more manageable. Chapter 4 discusses moment indexes in more detail.

An equipment list is furnished with the aircraft, which specifies all the required equipment, and all equipment approved for installation in the aircraft. The weight and arm of each item is included on the list, and all equipment installed when the aircraft left the factory is checked.

When an aircraft mechanic or repairman adds or removes any item on the equipment list, he or she must change the weight and balance record to indicate the new empty weight and EWCG, and the equipment list is revised to show which equipment is actually installed. Figure 2-22 is an excerpt from a comprehensive equipment list that includes all of the items of equipment approved for this particular model of aircraft. The POH for each individual aircraft includes an aircraft specific equipment list of the items from this master list. When any item is added to or removed from the aircraft, its weight and arm are determined in the equipment list and used to update the weight and balance record.

The POH/AFM also contains CG moment envelopes and loading graphs. Examples of the use of these handy graphs are given in chapter 4.

Weight and Balance Data

Aircraft Serial No. 18259080 **F.A.A. Registration No. N42565** **Date: 4-22-05**

Item	Weight (lbs) X	C.G. Arm (in) =	Moment (lb-in)
Standard empty weight	1,876	36.1	67,798.6
Optional equipment	1.2	13.9	16.7
Special installation	6.2	41.5	257.3
Paint	—	—	—
Unusable fuel	30.0	46.0	1,380
Basic empty weight	1,913.4		69,452.6

Figure 2-21. *Typical weight and balance data for 14 CFR part 23 airplane.*

Comprehensive Equipment List

The following figure (Figure 6-9) is a comprehensive list of all Cessna equipment which is available for the Model 182S airplane. It should not be confused with the airplane-specific equipment list. An airplane-specific list is provided with each individual airplane at delivery, and is typically inserted at the rear of this Pilot's Operating Handbook. The following comprehensive equipment list and the airplane-specific list have a similar order of listing.

The comprehensive equipment list provides the following information in column form:

In the **Item No** column, each item is assigned a coded number. The first two digits of the code represent the assignment of item within the ATA Specification 100 breakdown (Chapter 11 for Placards, Chapter 21 for Air Conditioning, Chapter 77 for Engine Indicating, etc…). These assignments also correspond to the Maintenance Manual chapter breakdown for the airplane. After the first two digits (and hyphen), items receive a unique sequence number (01, 02, 03, etc…). After the sequence number (and hyphen), a suffix letter is assigned to identify equipment as a required item, a standard item or an optional item. Suffix letters are as follows:

–R = required items or equipment for FAA certification
–S = standard equipment items
–O = optional equipment items replacing required or standard items
–A = optional equipment items which are in addition to required or standard items

In the **Equipment List Description** column, each item is assigned a descriptive name to help identify its function.

In the **Ref Drawing** column, a drawing number is provided which corresponds to the item.

Note

If additional equipment is to be installed, it must be done in accordance with the reference drawing, service bulletin or a separate FAA approval.

In the **Wt Lbs** and **Arm Ins** columns, information is provided on the weight (in pounds) and arm (in inches) of the equipment item.

Notes

Unless otherwise indicated, true values (not net change values) for the weight and arm are shown. Positive arms are distances aft of the airplane datum; negative arms are distances forward of the datum.

Asterisks (*) in the weight and arm column indicate complete assembly installations. Some major components of the assembly are listed on the lines immediately following. The sum of these major components does not necessarily equal the complete assembly installation.

Figure 2-22. *Excerpt from a typical comprehensive equipment list.*

Item No	Equipment List Description	Ref Drawing	Wt (lbs.)	Arm (ins.)
24-04-S	Basic Avionics Kit Installation		4.3*	55.5*
	- Support Straps Installation		0.1	10.0
	- Avionics Cooling Fan Installation		1.6	3.0
	- Avionics Ground Installations		0.1	41.0
	- Circuit Breaker Panel Installation		1.5	16.5
	- Microphone Installation		0.2	18.5
	- Omni Antenna Installation		0.5	252.1
	- Omni Antenna Cable Assembly Installation		0.3	248.0
	Chapter 25 – Equipment/Furnishings			
25-01-R	Seat, Pilot, Adjustable		33.8	41.5
25-02-S	Seat, Copilot, Adjustment		33.8	41.5
25-03-S	Seat, Rear, Two Piece Back Cushion		50.0	82.0
25-04-R	Seat Belt and Shoulder Harness, Inertia Reel, Pilot and Copilot		5.2	50.3
25-05-S	Seat Belt and Shoulder Harness, Inertia Reel, Rear Seat		5.2	87.8
25-06-S	Sun Visors (Set of 2)		1.2	33.0
25-07-S	Baggage Retaining Net		0.5	108.0
25-08-S	Cargo Tie Down Rings (10 Tie Downs)		0.4	108.0
25-09-S	Pilot's Operating Checklist (Stowed in Instrument Panel Map Case)		0.3	15.0
25-10-R	Pilot's Operating Handbook and FAA Approved Airplane Flight Manual (Stowed in Pilot's Seat Back)		1.2	61.5
25-11-S	Fuel Sampling Cup		0.1	14.3
25-12-S	Tow Bar, Nose Gear (Stowed)		1.7	108.0
25-13-S	Emergency Locator Transmitter Installation		2.2*	134.8*
	- ELT Transmitter		1.7	135.0
	- Antenna and Cable Assembly		0.4	133.0
	- Hardware		0.1	138.0
	Chapter 26 – Fire Protection			
26-01-S	Fire Extinguisher Installation		5.3*	29.0*
	- Fire Extinguisher		4.8	29.0
	- Mounting Clamp & Hardware		0.5	29.0
	Chapter 27 – Flight Controls			
27-01-S	Dual Controls Installation, Right Seat		6.3*	12.9*
	- Control Wheel, Copilot		2.0	26.0
	- Rudder and Brake Pedal Installation Copilot		4.3	6.8

Figure 2-22. *Excerpt from a typical comprehensive equipment list (continued).*

Weighing the Aircraft and Determining the Empty Weight-Center of Gravity

Chapter 2 explained the theory of weight and balance and gave examples of the way the center of gravity could be found for a board loaded with several weights. In this chapter, the practical aspects of weighing an airplane and locating its center of gravity are discussed. Formulas are introduced that allow the CG location to be measured in inches from various datum locations and in percentage of the mean aerodynamic chord.

Requirements

Weight and balance is of such vital importance that each mechanic or repairman maintaining an aircraft must be fully aware of his or her responsibility to provide the pilot with current and accurate information for the actual weight of the aircraft and the location of the center of gravity. The pilot in command has the responsibility to know the weight of the load, CG, maximum allowable weight, and CG limits of the aircraft.

The weight and balance report must include an equipment list showing weights and moment arms of all required and optional items of equipment included in the certificated empty weight.

When an aircraft has undergone extensive repair or major alteration, it should be reweighed and a new weight and balance record started. The A&P mechanic or the repairman responsible for the work must provide the pilot with current and accurate aircraft weight information and where the new EWCG is located.

Equipment for Weighing

There are two basic types of scales used to weigh aircraft: scales on which the aircraft is rolled so that the weight is taken at the wheels, and electronic load cells type where a pressure sensitive cell are placed between the aircraft jack and the jack pads on the aircraft.

Some aircraft are weighed with mechanical scales of the low-profile type similar to those shown in Figure 3-1.

Large aircraft, including heavy transports, are weighed by rolling them onto weighing platforms with electronic weighing cells that accurately measure the force applied by the weight of the aircraft.

Electronic load cells are used when the aircraft is weighed by raising it on jacks. The cells are placed between the jack and the jack pad on the aircraft, and the aircraft is raised on the jacks until the wheels or skids are off the floor and the aircraft is in a level flight attitude. The weight measured by each load cell is indicated on the control panel. [Figure 3-27]

Mechanical scales should be protected when they are not in use, and they must be periodically checked for accuracy by measuring a known weight. Electronic scales normally have a built-in calibration that allows them to be accurately zeroed before any load is applied.

Figure 3-1. *Low profile mechanical platform scales are used to weigh some aircraft. One scale is placed under each wheel.*

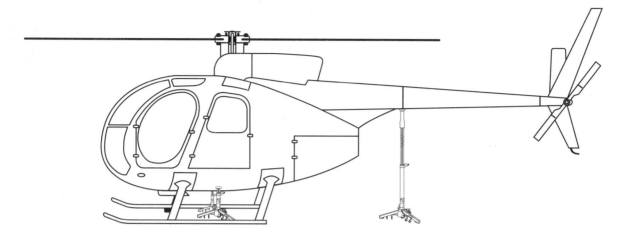

Figure 3-2. *Electronic load cell scale. A load cell is placed at each jack point.*

Preparation for Weighing

The major considerations in preparing an aircraft for weighing are discussed below.

Weigh Clean Aircraft Inside Hangar

The aircraft should be weighed inside a hangar where wind cannot blow over the surface and cause fluctuating or false scale readings.

The aircraft should be clean inside and out, with special attention paid to the bilge area to be sure no water or debris is trapped there, and the outside of the aircraft should be as free as possible of all mud and dirt.

Equipment List

All of the required equipment must be properly installed, and there should be no equipment installed that is not included in the equipment list. If such equipment is installed, the weight and balance record must be corrected to indicate it.

Ballast

All required permanent ballast must be properly secured in place and all temporary ballast must be removed.

Draining the Fuel

Drain fuel from the tanks in the manner specified by the aircraft manufacturer. If there are no specific instructions, drain the fuel until the fuel quantity gauges read empty when the aircraft is in level-flight attitude. Any fuel remaining in the system is considered residual, or unusable fuel and is part of the aircraft empty weight.

If it is not feasible to drain the fuel, the tanks can be topped off to be sure of the quantity they contain and the aircraft weighed with full fuel. After weighing is complete, the weight of the fuel and its moment are subtracted from those of the aircraft as weighed. To correct the empty weight for the residual fuel, add its weight and moment. The amount of residual fuel and its arm are normally found in NOTE 1 in the section of the TCDS, "Data pertaining to all Models." See "Fuel Capacity" on page 2-10.

When computing the weight of the fuel, for example a tank full of jet fuel, measure its specific gravity (sg) with a hydrometer and multiply it by 8.345 (the nominal weight of 1 gallon of pure water whose s.g. is 1.0). If the ambient temperature is high and the jet fuel in the tank is hot enough for its specific gravity to reach 0.81 rather than its nominal s.g. of 0.82, the fuel will actually weigh 6.76 pounds per gallon rather than its normal weight of 6.84 pounds per gallon. The standard weight of aviation gasoline (Avgas) is 6 pounds per gallon.

Oil

The empty weight for aircraft certificated under the CAR, part 3 does not include the engine lubricating oil. The oil must either be drained before the aircraft is weighed, or its weight must be subtracted from the scale readings to determine the empty weight. To weigh an aircraft that does not include the engine lubricating oil as part of the empty weight, place it in level flight attitude, then open the drain valves and allow all of the oil that is able, to drain out. Any remaining is undrainable oil, and is part of the empty weight. Aircraft certificated under 14 CFR parts 23 and 25 include full oil as part of the empty weight. If it is impractical to drain the oil, the reservoir can be filled to the specified level and the weight of the oil computed at 7.5 pounds per gallon. Then its weight and moment are subtracted from the weight and moment of the aircraft as weighed. The amount and arm of the undrainable oil are found in NOTE 1 of the TCDS, and this must be added to the empty weight.

Other Fluids

The hydraulic fluid reservoir and all other reservoirs containing fluids required for normal operation of the aircraft should be full. Fluids not considered to be part of the empty weight of the aircraft are potable (drinkable) water, lavatory precharge water, and water for injection into the engines.

Configuration of the Aircraft

Consult the aircraft service manual regarding position of the landing gear shock struts and the control surfaces for weighing; when weighing a helicopter, the main rotor must be in its correct position.

Jacking the Aircraft

Aircraft are often weighed by rolling them onto ramps in which load cells are embedded. This eliminates the problems associated with jacking the aircraft off the ground. However, many aircraft are weighed by jacking the aircraft up and then lowering them onto scales or load cells.

Extra care must be used when raising an aircraft on jacks for weighing. If the aircraft has spring steel landing gear and it is jacked at the wheel, the landing gear will slide inward as the weight is taken off of the tire, and care must be taken to prevent the jack from tipping over.

For some aircraft, stress panels or plates must be installed before they are raised with wing jacks, to distribute the weight over the jack pad. Be sure to follow the recommendations of the aircraft manufacturer in detail anytime an aircraft is jacked. When using two wing jacks, take special care to raise them simultaneously, keeping the aircraft so it will not slip off the jacks. As the jacks are raised, keep the safety collars screwed down against the jack cylinder to prevent the aircraft from tilting if one of the jacks should lose hydraulic pressure.

Leveling the Aircraft

When an aircraft is weighed, it must be in its level flight attitude so that all of the components will be at their correct distance from the datum. This attitude is determined by information in the TCDS. Some aircraft require a plumb line to be dropped from a specified location so that the point of the weight, the bob, hangs directly above an identifiable point. Others specify that a spirit level be placed across two leveling lugs, often special screws on the outside of the fuselage. Other aircraft call for a spirit level to be placed on the upper door sill.

Lateral level is not specified for all light aircraft, but provisions are normally made on helicopters for determining both longitudinal and lateral level. This may be done by built-in leveling indicators, or by a plumb bob that shows the conditions of both longitudinal and lateral level.

The actual adjustments to level the aircraft using load cells are made with the jacks. When weighing from the wheels, leveling is normally done by adjusting the air pressure in the nose wheel shock strut.

Safety Considerations

Special precautions must be taken when raising an aircraft on jacks.

1. Stress plates must be installed under the jack pads if the manufacturer specifies them.

2. If anyone is required to be in the aircraft while it is being jacked, there must be no movement.

3. The jacks must be straight under the jack pads before beginning to raise the aircraft.

4. All jacks must be raised simultaneously and the safety devices are against the jack cylinder to prevent the aircraft tipping if any jack should lose pressure. Not all jacks have screw down collars, some use drop pins or friction locks.

Determining the Center of Gravity

When the aircraft is in its level flight attitude, drop a plumb line from the datum and make a mark on the hangar floor below the tip of the bob. Draw a chalk line through this point parallel to the longitudinal axis of the aircraft. Then draw lateral lines between the actual weighting points for the main wheels, and make a mark along the longitudinal line at the weighing point for the nose wheel or the tail wheel. These lines and marks on the floor allow you to make accurate measurements between the datum and the weighting points to determine their arms.

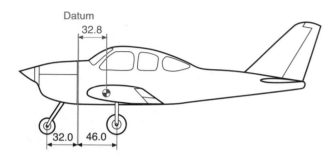

Figure 3-3. *The datum is located at the firewall.*

Determine the CG by adding the weight and moment of each weighing point to determine the total weight and total moment. Then divide the total moment by the total weight to determine the CG relative to the datum.

As an example of locating the CG with respect to the datum, which in this case is the firewall, consider the tricycle landing gear airplane in Figures 3-3 and 3-4.

When the airplane is on the scales with the parking brakes off, place chocks around the wheels to keep the airplane from rolling. Subtract the weight of the chocks, called tare weight, from the scale reading to determine the net weight at each weighing point. Multiply each net weight by its arm to determine its moment, and then determine the total weight and total moment. The CG is determined by dividing the total moment by the total weight.

$$CG = \frac{\text{Total moment}}{\text{Total weight}}$$
$$= \frac{65,756}{2,006}$$
$$= 32.8 \text{ inches behind the datum}$$

The airplane in Figures 3-3 and 3-4 has a net weight of 2,006 pounds, and its CG is 32.8 inches behind the datum.

Two Ways to Express CG Location

The location of the CG may be expressed in terms of inches from a datum specified by the aircraft manufacturer, or as a percentage of the MAC. The location of the leading edge of the MAC, the leading edge mean aerodynamic cord (LEMAC), is a specified number of inches from the datum.

Weighing Point	Scale Reading (lb)	Tare (lb)	Net Weight (lb)	Arm (in)	Moment (lb-in)	CG
Right side	846	16	830	46.0	38,180	
Left side	852	16	836	46.0	38,456	
Nose	348	8	340	−32.0	−10,880	
Total			2,006		65,756	32.8

Figure 3-4. *Locating the CG of an airplane relative to the datum, which is located at the firewall. See Figure 3-3.*

Empty-Weight Center of Gravity Formulas

A chart like the one in Figure 3-4 helps visualize the weights, arms, and moments when solving an EWCG problem, but it is quicker to determine the EWCG by using formulas and an electronic calculator. The use of a calculator for solving these problems is described in chapter 8.

There are four possible conditions and their formulas that relate the location of CG to the datum. Notice that the formula for each condition first determines the moment of the nose $\left(\frac{F \times L}{W}\right)$ wheel or tail $\left(\frac{R \times L}{W}\right)$ wheel and then divides it by the total weight of the airplane. The arm thus determined is then added to or subtracted from the distance between the main wheels and the datum, distance D.

Nose wheel airplanes with datum forward of the main wheels.

$$CG = D - \left(\frac{F \times L}{W}\right)$$

Nose wheel airplanes with the datum aft of the main wheels.

$$CG = -\left(D + \frac{F \times L}{W}\right)$$

Tail wheel airplanes with the datum forward of the main wheels.

$$CG = D + \left(\frac{R \times L}{W}\right)$$

Tail wheel airplanes with the datum aft of the main wheels.

$$CG = -D + \left(\frac{R \times L}{W}\right)$$

Datum Forward of the Airplane - Nose Wheel Landing Gear

The datum of the airplane in Figure 3-5 is 100 inches forward of the leading edge of the wing root, or 128 inches forward of the main-wheel weighing points. This is distance (D). The weight of the nose wheel (F) is 340 pounds, and the distance between main wheels and nose wheel (L) is 78 inches. The total weight of the airplane (W) is 2,006 pounds.

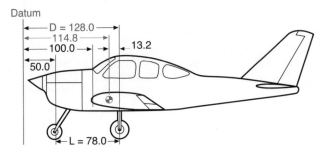

Figure 3-5. *The datum is 100 inches forward of the wing root leading edge.*

Determine the CG by using this formula:

$$CG = D - \left(\frac{F \times L}{W}\right)$$
$$= 128 - \left(\frac{340 \times 78}{2,006}\right)$$
$$= 114.8$$

The CG is 114.8 inches aft of the datum. This is 13.2 inches forward of the main-wheel weighing points which proves the location of the datum has no effect on the location of the CG so long as all measurements are made from the same location.

Datum Aft of the Main Wheels - Nose Wheel Landing Gear

The datum of some aircraft may be located aft of the main wheels. The airplane in this example is the same one just discussed, but the datum is at the intersection of the trailing edge of the wing with the fuselage.

The distance (D) between the datum of the airplane in Figure 3-6 and the main-wheel weighing points is 75 inches, the weight of the nose wheel (F) is 340 pounds, and the distance between main wheels and nose wheel (L) is 78 inches. The total net weight of the airplane (W) is 2,006 pounds.

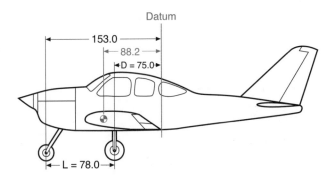

Figure 3-6. *The datum is aft of the main wheels at the wing trailing edge.*

The location of the CG may be determined by using this formula:

$$CG = -\left(D + \frac{F \times L}{W}\right)$$
$$= -\left(75 + \frac{340 \times 78}{2,006}\right)$$
$$= -88.2$$

The CG location is a negative value, which means it is 88.2 inches forward of the datum. This places it 13.2 inches forward of the main wheels, exactly the same location as it was when it was measured from other datum locations.

Location of Datum

It makes no difference where the datum is located as long as all measurements are made from the same location.

Datum Forward of the Main Wheels- Tail Wheel Landing Gear

Locating the CG of a tail wheel airplane is done in the same way as locating it for a nose wheel airplane except the formulas use $\left(\frac{R \times L}{W}\right)$ rather than $\left(\frac{F \times L}{W}\right)$.

The distance (D) between the datum of the airplane in Figure 3-7 and the main-gear weighing points is 7.5 inches, the weight of the tail wheel (R) is 67 pounds, and the distance (L) between the main-wheel and the tail wheel weighing points is 222 inches. The total weight of the airplane (W) is 1,218 pounds.

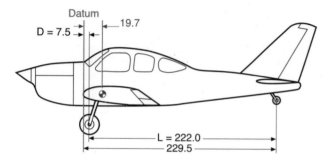

Figure 3-7. *The datum of this tail wheel airplane is the wing root leading edge.*

Determine the CG by using this formula:

$$CG = D + \left(\frac{R \times L}{W}\right)$$
$$= 7.5 + \left(\frac{67 \times 222}{1,218}\right)$$
$$= 19.7$$

The CG is 19.7 inches behind the datum.

Datum Aft of the Main Wheels - Tail Wheel Landing Gear

The datum of the airplane in Figure 3-8 is located at the intersection of the wing root trailing edge and the fuselage. This places the arm of the main gear (D) at -80 inches. The net weight of the tail wheel (R) is 67 pounds, the distance between the main wheels and the tail wheel (L) is 222 inches, and the total net weight (W) of the airplane is 1,218 pounds.

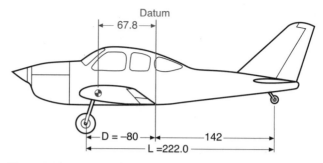

Figure 3-8. *The datum is aft of the main wheels, at the intersection of the wing trailing edge and the fuselage.*

Since the datum is aft of the main wheels, use the formula:

$$CG = -D + \left(\frac{R \times L}{W}\right)$$
$$= -80 + \left(\frac{67 \times 222}{1,218}\right)$$
$$= -67.8$$

The CG is 67.8 inches forward of the datum, or 12.2 inches aft of the main-gear weighing points. The CG is in exactly the same location relative to the main wheels, regardless of where the datum is located.

Location with Respect to the Mean Aerodynamic Chord

The aircraft mechanic or repairman is primarily concerned with the location of the CG relative to the datum, an identifiable physical location from which measurements can be made. But because the physical chord of a wing that does not have a strictly rectangular plan form is difficult to measure, wings such as tapered wings express the allowable CG range in percentage of mean aerodynamic chord (MAC). The allowable CG range is expressed in percentages of the MAC. The MAC, as seen in Figure 3-9, is the chord of an imaginary airfoil that has all of the aerodynamic characteristics of the actual airfoil. It can also be thought of as the chord drawn through the geographic center of the plan area of the wing.

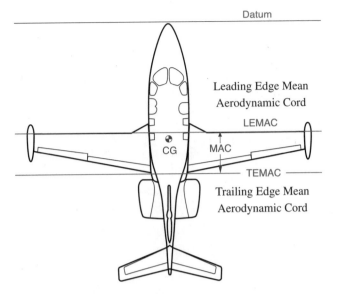

The CG of the airplane is located at 27.4% MAC.

It is sometimes necessary to determine the location of the CG in inches from the datum when its location in %MAC is known.

> The CG of the airplane is located at
> 27.4% MAC
> MAC = 206 - 144 = 62
> LEMAC = station 144

Determine the location of the CG in inches from the datum by using this formula:

$$\text{CG inches from datum} = \text{LEMAC} + \frac{\text{MAC} \times \text{CG \% MAC}}{100}$$

$$= 144 + \frac{62 \times 27.4}{100}$$

$$= 160.9$$

The CG of this airplane is located at station 160.9 inches aft of the datum. It is important for longitudinal stability that the CG be located ahead of the center of lift of a wing. Since the center of lift is expressed as a percentage of the MAC, the location of the CG is expressed in the same terms.

Figure 3-9. *The MAC is the chord drawn through the geographic center of the plan area of the wing.*

The relative positions of the CG and the aerodynamic center of lift of the wing have critical effects on the flight characteristics of the aircraft.

Consequently, relating the CG location to the chord of the wing is convenient from a design and operations standpoint. Normally, an aircraft will have acceptable flight characteristics if the CG is located somewhere near the 25 percent average chord point. This means the CG is located one-fourth of the total distance back from the leading edge of the wing section. Such a location will place the CG forward of the aerodynamic center for most airfoils.

In order to relate the percent MAC to the datum, all weight and balance information includes two items: the length of MAC in inches and the location of the leading edge of MAC (LEMAC) in inches from the datum.

The weight and balance data of the airplane in Figure 3-10 states that the MAC is from stations 144 to 206 and the CG is located at station 161.

> MAC = 206" - 144" = 62" inches
> LEMAC = station 144
> CG is 17 inches behind LEMAC
> (160 - 144 = 17.0 inches)

The location of the CG expressed in percentage of MAC is determined using this formula:

$$\text{CG in \% MAC} = \frac{\text{Distance aft of LEMAC} \times 100}{\text{MAC}}$$

$$= \frac{17 \times 100}{62}$$

$$= 27.4$$

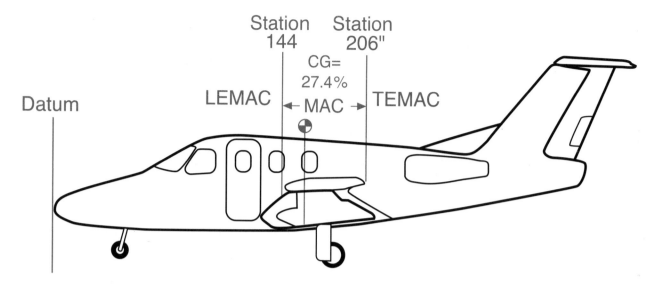

Figure 3-10. *Aircraft weight and balance calculation diagram.*

Small Fixed Wing Aircraft Operational Weight and Balance Computations

Weight and balance data allows the pilot to determine the loaded weight of the aircraft and determine whether or not the loaded CG is within the allowable range for the weight. See Figure 4-1 for an example of the data necessary for these calculations.

Airplane basic empty weight	1,874.0 lbs, EWCG +36.1
CG range	(+40.9) to (+46.0) at 3,100 lbs (+33.0) to (+46.0) at 2,250 lbs or less Straight line variation between points given
Empty weight CG range	None
Maximum weight	3,100 lbs takeoff/flight 2,950 lbs landing
No. of seats	4 (2 front at +37.0) (2 rear at +74.0)
Maximum baggage	160 lbs Area A (100 lbs at +97.0) Area B (60 lbs at +116.0)
Fuel capacity	92 gal (88 gal usable); two 46 gal integral tanks in wings at +46.6 See NOTE 1 for data on unusable fuel.
Oil capacity	12 qt (–15)

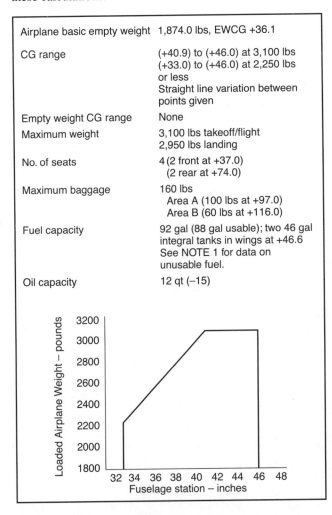

Figure 4-1. *Weight and balance data needed to determine proper loading of a small fixed wing aircraft.*

Determining the Loaded Weight and CG

An important part of preflight planning is to determine that the aircraft is loaded so its weight and CG location are within the allowable limits. [Figure 4-2] There are two ways of doing this: by the computational method using weight, arms, and moments; and by the loading graph method, using weight and moment indexes.

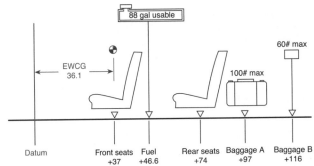

Figure4-2. *Airplane loading diagram.*

Computational Method

The computational method uses weights, arms, and moments. It relates the total weight and CG location to a CG limits chart similar to those included in the TCDS and the POH/AFM.

A worksheet such as the one in Figure 4-3 provides space for all of the pertinent weight, CG, and moment along with the arms of the seats, fuel, and baggage areas.

Item	Weight (3,100 max.)	Arm (inches)	Moment (lb-in)	CG (in/datum)
Airplane (BEW)	1,874	36.1	67,651.4	
Front seats		37		
Rear seats		74		
Fuel (88 gal usable)		46.6		
Baggage A (100 max.)		97		
Baggage B (60 max.)		116		

Figure 4-3. *Blank weight and balance worksheet.*

Item	Weight (3,100 max.)	Arm (inches)	Moment (lb-in)	CG (in/datum)
Airplane (BEW)	1,874	36.1	67,651.4	
Front seats	300	37	11,100	
Rear seats	175	74	12,950	
Fuel (88 gal usable)	528	46.6	24,604.8	
Baggage A (100 max.)	100	97	9,700	
Baggage B (60 max.)	50	116	5,800	
	3,027		131,806.2	+ 43.54

Figure 4-4. *Completed weight and balance worksheet.*

When planning the flight, fill in the blanks in the worksheet with the specific data for the flight. [Figure 4-4]

Pilot...120 lbs
Front seat passenger....................180 lbs
Rear seat passenger.....................175 lbs
Fuel 88 gal ....528 lbs
Baggage A ....100 lbs
Baggage B....................................50 lbs

Determine the moment of each item by multiplying its weight by its arm. Then determine the total weight and the sum of the moments. Divide the total moment by the total weight to determine the CG in inches from the datum. The total weight is 3,027 pounds and the CG is 43.54 inches aft of the datum.

To determine that the airplane is properly loaded for this flight, use the CG limits envelope in Figure 4-5 (which is typical of those found in the POH/AFM). Draw a line vertically upward from the CG of 43.54 inches, and one horizontally to the right from the loaded weight of 3,027 pounds. These lines cross inside the envelope, which shows the airplane is properly loaded for takeoff, but 77 pounds overweight for landing.

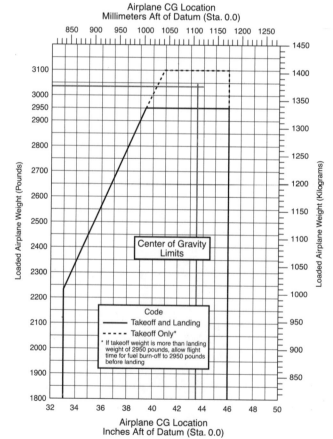

Figure 4-5. *Center of gravity limits chart from a typical POH.*

Loading Graph Method

Everything possible is done to make flying safe, and one expedient method is the use of charts and graphs from the POH/AFM to simplify and speed up the preflight weight and balance computation. Some use a loading graph and moment indexes rather than the arms and moments. These charts eliminate the need for calculating the moments and thus make computations quicker and easier. [Figure 4-5]

Moment Indexes

Moments determined by multiplying the weight of each component by its arm result in large numbers that are awkward to handle and can become a source of mathematical error. To eliminate these large numbers, moment indexes are used. The moment is divided by a reduction factor such as 100 or 1,000 to get the moment index. The loading graph provides the moment index for each component, so you can avoid mathematical calculations. The CG envelope uses moment indexes rather than arms and moments.

CG limits envelope: is the enclosed area on a graph of the airplane loaded weight and the CG location. If lines drawn from the weight and CG cross within this envelope, the airplane is properly loaded.

Loading Graph

Figure 4-6 is a typical loading graph taken from the POH of a modern four-place airplane. It is a graph of load weight and load moment indexes. Diagonal lines for each item relate the weight to the moment index without having to use mathematical calculations.

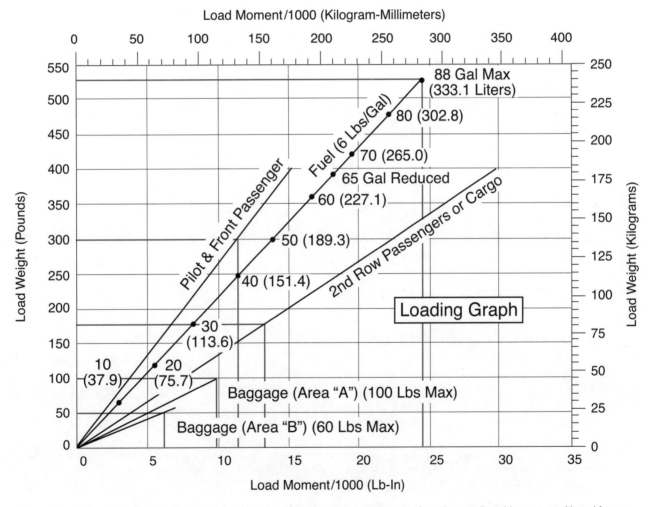

Note: Line representing adjustable seats shows pilot and front seat passenger center of gravity on adjustable seats positioned for an average occupant. Refer to the Loading Arrangements diagram for forward and aft limits of occupant CG range.

Figure 4-6. *Typical loading graph.*

Compute Weight and Balance Using the Loading Graph

To compute the weight and balance using the loading graph in Figure 4-6, make a loading schedule chart like the one in Figure 4-7.

In Figure 4-6, follow the horizontal line for 300 pounds load weight to the right until it intersects the diagonal line for pilot and front passenger. From this point, drop a line vertically to the load moment index along the bottom to determine the load moment for the front seat occupants. This is 11.1 lb-in/1,000. Record it in the loading schedule chart.

Determine the load moment for the 175 pounds of rear seat occupants along the diagonal for second row passengers or cargo. This is 12.9; record it in the loading schedule chart.

Item	Weight	Moment/1000
Airplane (BEW)	1,874	67.7
Front seat	300	11.1
Rear seat	175	12.9
Fuel	528	24.6
Baggage A	100	9.7
Baggage B	50	5.8
Total	3,027	131.8

Figure 4-7. *Loading schedule chart.*

Determine the load moment for the fuel and the baggage in areas A and B in the same way and enter them all in the loading schedule chart. The maximum fuel is marked on the diagonal line for fuel in terms of gallons or liters. The maximum is 88 gallons of usable fuel. The total capacity is 92 gallons, but 4 gallons are unusable and have already been included in the empty weight of the aircraft. The weight of 88 gallons of gasoline is 528 pounds and its moment index is 24.6. The 100 pounds of baggage in area A has a moment index of 9.7 and the 50 pounds in area B has an index of 5.8. Enter all of these weights and moment indexes in the loading schedule chart and add all of the weights and moment indexes to determine the totals. Transfer these values to the CG moment envelope in Figure 4-8.

The CG moment envelope is an enclosed area on a graph of the airplane loaded weight and loaded moment. If lines drawn from the weight and loaded moment cross within this envelope, the airplane is properly loaded.

The loading schedule shows that the total weight of the loaded aircraft is 3,027 pounds, and the loaded airplane moment/1,000 is 131.8.

Draw a line vertically upward from 131.8 on the horizontal index at the bottom of the chart, and a horizontal line from 3,027 pounds in the left-hand vertical index. These lines intersect within the dashed area, which shows that the aircraft is loaded properly for takeoff, but it is too heavy for landing.

If the aircraft had to return for landing, it would have to fly long enough to burn off 77 pounds (slightly less than 13 gallons) of fuel to reduce its weight to the amount allowed for landing.

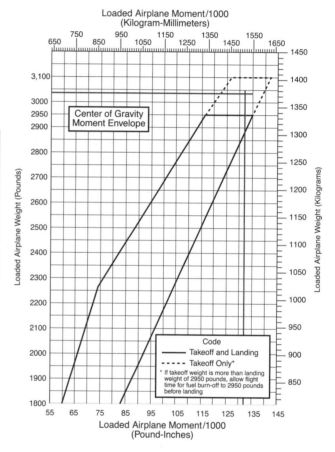

Figure 4-8. *CG moment envelope.*

Multiengine Airplane Weight and Balance Computations

Weight and balance computations for small multiengine airplanes are similar to those discussed for single-engine airplanes. See Figure 4-9 for an example of weight and balance data for a typical light twin-engine airplane.

Datum — Forward face of fuselage bulkhead ahead of rudder pedals
Seats
   2 at 37.0
   2 at 75.0
   1 at 113.0 — Weight limit 200 lbs
Fuel
   213.4 gal (2 wing tanks, 105.0 gal each 103.0 gal usable at +61.0)
   Undrainable fuel — 1.6 lbs at +62
Oil
   24 quarts (12 quarts in each engine) — −3.3
Baggage
   Forward 100# limit — −15
   Aft 200# limit — +113
CG Range
   (+38) to (+43.1) at 5,200 lbs
   (+43.6) at 4,800 lbs
   (+32) to (+43.6) at 4,300 lbs or less
   Straight line variation between points given
Engines (2) 240 horsepower horizontally opposed engines
   Fuel burn — 24 gph for 65% cruise at 175 knot
                29 gph for 75% cruise at 180 knot

Figure 4-9. *Typical weight and balance data for a light twin-engine airplane.*

The airplane in this example was weighed to determine its basic empty weight and EWCG. The weighing conditions and results are:

    Fuel drained -
    Oil full -
    Right wheel scales -1,084 lbs, tare 8 lbs
    Left wheel scales - 1,148 lbs, tare 8 lbs
    Nose wheel scales - 1,202 lbs, tare 14 lbs

Determine the Loaded CG

Beginning with the basic empty weight and EWCG and using a chart such as the one in Figure 4-11, the loaded weight and CG of the aircraft can be determined. [Figure 4-10]

The aircraft is loaded as shown here:

    Fuel (140 gal).................... 840 lbs
    Front seats....................... 320 lbs
    Row 2 seats .... 310 lbs
    Fwd. baggage................... 100 lbs
    Aft. baggage...................... 90 lbs

Chart Method Using Weight, Arm, and Moments

Make a chart showing the weight, arm, and moments of the airplane and its load.

Item	Weight pounds (5,200 max.)	Arm (inches)	Moment (lb-in)	CG
Aircraft	3,404	35.28	120,093	
Fuel (140 gal)	840	61.0	51,240	
Front seat	320	37.0	11,840	
Row 2 seats	310	75.0	23,250	
Fwd. baggage	100	−15	−1,500	
Aft baggage	90	113	10,170	
Total	5,064		215,093	42.47

Figure 4-11. *Determining the loaded center of gravity of the airplane in Figure 4-10.*

The loaded weight for this flight is 5,064 pounds, and the CG is located at 42.47 inches aft of the datum.

To determine that the weight and CG are within the allowable range, refer to the CG range chart of Figure 4-12. Draw a line vertically upward from 42.47 inches from the datum and one horizontally from 5,064 pounds. These lines cross inside the envelope, showing that the airplane is properly loaded.

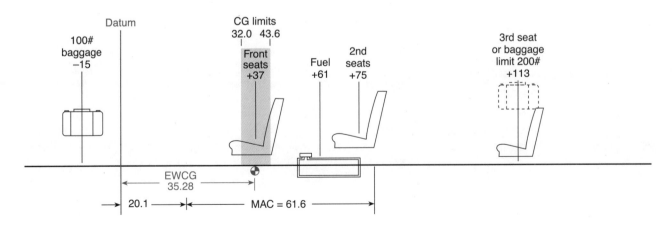

Figure 4-10. *Twin-engine airplane weight and balance diagram.*

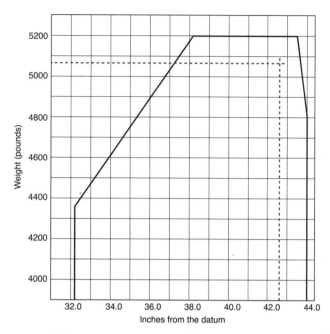

Figure 4-12. *Center of gravity range chart.*

Determining the CG in Percent of MAC

Refer again to Figures 4-10 and 4-11.

> The loaded CG is 42.47 inches aft of the datum.
> The MAC is 61.6 inches long.
> The LEMAC is located at station 20.1.
> The CG is 42.47 - 20.1 = 22.37 inches aft of LEMAC.

Use this formula:

$$CG \text{ in } \% \text{ MAC} = \frac{CG \text{ in inches from LEMAC} \times 100}{MAC}$$

$$= \frac{22.37 \times 100}{61.6}$$

$$= 36.3\% \text{ MAC}$$

The loaded CG is located at 36.3% of the mean aerodynamic chord.

The Chart Method Using Weight and Moment Indexes

As mentioned in the previous chapter, anything that can be done to make careful preflight planning easier makes flying safer. Many manufacturers furnish charts in the POH/AFM that use weight and moment indexes rather than weight, arm, and moments. They further help reduce errors by including tables of moment indexes for the various weights.

Consider the loading for this particular flight:

> Cruise fuel flow = 16 gallons per hour
> Estimated time en route = 2 hours 10 minutes.
> Reserve fuel = 45 minutes = 12 gallons
> Total required fuel = 47 gallons

The pilot completes a chart like the one in Figure 4-13 using moment indexes from tables in figure 4-14 through 4-16.

The moments/100 in the index column are found in the charts in Figure 4-14 through 4-16. If the exact weight is not in the chart, interpolate between the weights that are included. When a weight is greater than any of those shown in the charts, add the moment indexes for a combination of weights to get that which is desired. For example, to get the moments/100 for the 320 pounds in the front seats, add the moment index for 100 pounds (105) to that for 220 pounds (231). This gives the moment index of 336 for 320 pounds in the front seats.

Use the moment limits vs. weight envelope in Figure 4-17 on page 4-8 to determine if the weight and balance conditions will be within allowable limits for both takeoff and landing at the destination.

The Moment limits vs. Weight envelope is an enclosed area on a graph of three parameters. The diagonal line representing the moment/100 crosses the horizontal line representing the weight at the vertical line representing the CG location in inches aft of the datum. When the lines cross inside the envelope, the aircraft is loaded within its weight and CG limits.

> Takeoff - 3,781 lbs and 4,310
> moment/100
>
> Landing - 3,571 lbs and 4,050
> moment/100

Locate the moment/100 diagonal line for 4,310 and follow it down until it crosses the horizontal line for 3,781 pounds. These lines cross inside the envelope at the vertical line for a CG location of 114 inches aft of the datum.

The maximum allowable takeoff weight is 3,900 pounds, and this airplane weighs 3,781 pounds. The CG limits for 3,781 pounds are 109.8 to 117.5. The CG of 114 inches falls within these allowable limits.

Weight and Balance Loading Form

Model _____   Date _____

Serial Number_____   Reg. Number _____

Item	Pounds (3,900 max.)	Index Moment/100
Airplane basic empty weight		
Front seat occupants	320	336
Row 2 seats	290	412
Baggage (200# max.)	90	150
Sub Total		
Zero fuel condition (3,500 max.)	3,325	3,762
Fuel loading – gallons _80_	480	562
Sub Total		
Ramp condition	3,805	4,324
*Less fuel for start, taxi, and takeoff	-24	-28
Sub Total		
Takeoff condition	3,781	4,296
Less fuel to destination – gallons _35_	-210	-246
Landing condition	3,571	4,050

* Fuel for start, taxi, and takeoff is normally 24 pounds at a moment index of 28.

Figure 4-13. *Typical weight and balance loading form.*

Occupants Moments/100		
Weight	**Front seats Arm +105**	**Row 2 seats Arm +142**
100	105	142
110	116	156
120	126	170
130	137	185
140	147	199
150	158	213
160	168	227
170	179	241
180	189	256
190	200	270
200	210	284
210	221	298
220	231	312
230	242	327
240	252	341
250	263	355

Figure 4-14. *Weight and moment index for occupants.*

Baggage Moments/100	
Weight	**Arm 167**
10	17
20	33
30	50
40	67
50	84
60	100
70	117
80	134
90	150
100	167
110	184
120	200
130	217
140	234
150	251
160	267
170	284
180	301
190	317
200	334

Figure 4-15. *Weight and moment index for baggage.*

Usable Fuel – Arm +117		
Gallons	Pounds	Moment/100
10	60	70
20	120	140
30	180	211
40	240	281
50	300	351
60	360	421
70	420	491
80	480	562
90	540	632
100	600	702

Figure 4-16. *Weight and moment index for fuel.*

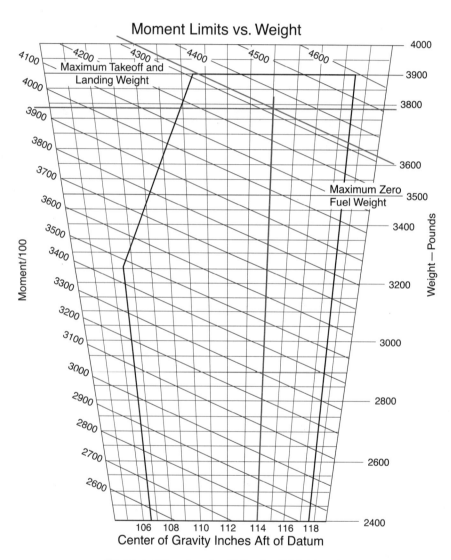

Figure 4-17. *Moment limits vs. weight envelope.*

Envelope Based on the Following Weight and
Center of Gravity Limit Data (Landing Gear Down)

Weight Condition	FWD CG Limit	Aft CG Limit
3,900 Pounds (Max Takeoff/Landing)	110.6	117.5
3,250 Pounds or Less	106.6	117.5

Center of Gravity Change after Repair or Alterations

The largest weight changes that occur during the lifetime of an aircraft are those caused by alterations and repairs. It is the responsibility of the aircraft mechanic or repairman doing the work to accurately document the weight change and record it in both the maintenance records and the POH/AFM.

Equipment List

A typical comprehensive equipment list is shown in Figure 2-22 on pages 2-12 and 2-13. The FAA considers addition or removal of equipment included in this list to be a minor alteration. The weights and arms are included with the items in the equipment list, and these minor alterations can be done and the aircraft approved for return to service by an appropriately rated aircraft mechanic or repairman. The only documentation required is an entry in the aircraft maintenance records and the appropriate change to the weight and balance record in the POH/AFM. [Figure 5-1]

Major Alteration and Repair

Within the following text, information concerning major repairs or major alterations does not apply to any aircraft within the light-sport category. This category of aircraft is not eligible for major repairs or alterations.

Any major alteration or repair requires the work to be done by an appropriately-rated aircraft mechanic or facility. The work must be checked for conformity to FAA-approved data and signed off by an aircraft mechanic holding an Inspection Authorization, or by an authorized agent of an appropriately rated FAA-approved repair station. A repair station record or FAA Form 337, Major Repair and Alteration, must be completed which describes the work. A dated and signed revision to the weight and balance record is made and kept with the maintenance records, and the airplane's new empty weight and empty weight arm or moment index are entered in the POH/AFM.

Weight and Balance Record
(Continuous history of changes in structure or equipment affecting weight and balance)

Airplane Model	*Cessna 182 L*		Serial Number	*18259080*		Page Number	*1*

Date	Item No.		Description of Article or Modification	Weight Change						Running Basic Empty Weight	
				Added (+)			Removed (-)				
	In	Out		Wt. (lb)	Arm (in)	Moment /1,000	Wt. (lb)	Arm (in)	Moment /1,000	Wt. (lb)	Moment /1,000
			AS DELIVERED							1,876	67.8
4-22-05			Alteration Per FAA Form 337 Dated 4-22-95	7.38		.346				1,883.4	68.1
	34-XX		Turn Coordinator				-2.5	15.0	-.037		
	34-XX		Directional Gyro				-3.12	13.5	-.042		
	22-XX		Auto Pilot System	13.0	32.7	.425					

Figure 5-1. *A typical Part 23 weight and balance record.*

Weight & Balance
Cessna 182L
N42565
S/N 18259080

Date: 04/22/95

Supersedes Computations of FAA Form 337, dated 10/02/90.

Removed the following equipment:

		Weight	Arm	Moment
1.	Turn Coordinator P/N C661003-0201	2.5 lbs	15	37.5
2.	Directional Gyro P/N 0706000	3.12 lbs	13.5	42.12
	TOTAL	5.62		79.62

	Weight	Arm	Moment
	1876.00	36.14	67798.64
	-5.62		-79.62
Aircraft after removal:	1870.38	36.20	67719.02

Installed the following equipment:

1. S-Tec 40 Autopilot system, includes Turn Coordinator and Directional Gyro.

	Weight	Arm	Moment
	13 lbs	32.7	425.13

	Weight	Arm	Moment
	1870.38	36.20	67719.02
	+13.00		+425.13
	1883.38	36.18	68144.15

*REVISED LICENSED EMPTY WEIGHT
NEW USEFUL LOAD 1216.62

Forward Check (Limit +33.0)

	Wt.	Arm	Moment
A/C Empty	1883.38	36.18	68144.15
Fwd. Seats	170.00	37.00	6290.00
Aft. Seats			
Fuel (min.)	115.00	48.00	5520.00
Baggage A			
Baggage B			
	2168.38	+36.87	79954.15

Rearward Check (Limit +46.0)

	Wt.	Arm	Moment
A/C Empty	1883.38	36.18	68144.15
Fwd. Seats	170.00	37.00	6290.00
Aft. Seats	340.00	74.00	25160.00
Fuel (max.)	528.00	48.20	25449.60
Baggage A	100.00	97.00	9700.00
Baggage B	60.00	116.00	6960.00
	3081.38	45.98	141703.75

Joseph P. Kline
Joseph P. Kline
A& P 123456789

Figure 5-2. *A typical CAR 3 airplane weight and balance revision record.*

Weight and Balance Revision Record

Aircraft manufacturers use different formats for their weight and balance data, but Figure 5-2 is typical of a weight and balance revision record. All weight and balance records should be kept with the other aircraft records. Each revision record should be identified by the date, the aircraft make, model, and serial number. The pages should be signed by the person making the revision and his or her certificate type and number must be included.

The computations for a weight and balance revision are included on a weight and balance revision form. The date those computations were made is shown in the upper right-hand corner of Figure 5-2. When this work is superseded, a notation must be made on the new weight and balance revision form, including a statement that these computations supersede the computations dated "XX/XX/XX."

Appropriate fore-and-aft extreme loading conditions should be investigated and the computations shown.

The weight and balance revision sheet should clearly show the revised empty weight, empty weight arm and/or moment index, and the new useful load.

Weight Changes Caused by a Repair or Alteration

A typical alteration might consist of removing two pieces of radio equipment from the instrument panel, and a power supply that was located in the baggage compartment behind the rear seat. In this example, these two pieces are replaced with a single lightweight, self-contained radio. At the same time, an old emergency locator transmitter (ELT) is removed from its mount near the tail, and a lighter weight unit is installed. A passenger seat is installed in the baggage compartment.

Computations Using Weight, Arm, and Moment

The first step in the weight and balance computation is to make a chart like the one in Figure 5-3, listing all of the items that are involved.

The new CG of 36.4 inches aft of the datum is determined by dividing the new moment by the new weight.

Item	Weight (lbs)	Arm (inches)	Moment (lb-in)	New CG
Airplane	1,876.0	36.1	67,723.6	
Radio removed	−12.2	15.8	−192.8	
Power supply removed	−9.2	95.0	−874.0	
ELT removed	−3.2	135.0	−432.0	
Radio installed	+8.4	14.6	+122.6	
ELT installed	+1.7	135.0	+229.5	
Passenger seat installed	+21.0	97.0	+2,037.0	
Total	1,882.5		68,613.9	+36.4

Figure 5-3. *Weight, arm, and moment changes caused by typical alteration.*

Computations Using Weight and Moment Indexes

If the weight and balance data uses moment indexes rather than arms and moments, this same alteration can be computed using a chart like the one shown on Figure 5-4.

Subtract the weight and moment indexes of all the removed equipment from the empty weight and moment index of the airplane. Add the weight and moment indexes of all equipment installed and determine the total weight and the total moment index. To determine the position of the new CG in inches aft of the datum, multiply the total moment index by 100 to get the moment, and divide this by the total weight to get the new CG.

Empty-Weight CG Range

The fuel tanks, seats, and baggage compartments of some aircraft are so located that changes in the fuel or occupant load have a very limited effect on the balance of the aircraft. Aircraft of such a configuration show an EWCG range in the TCDS. [Figure 5-5] If the EWCG is located within this range, it is impossible to legally load the aircraft so that its loaded CG will fall outside of its allowable range.

Empty-Weight CG Range	+12.5 to +16.2

Figure 5-5. *Typical notation in a TCDS when an aircraft has an empty-weight CG range.*

If the TCDS list an empty-weight CG range, and after the alteration is completed the EWCG falls within this range, then there is no need to compute a fore and aft check for adverse loading.

But if the TCDS lists the EWCG range as "None" (and most of them do), a check must be made to determine whether or not it is possible by any combination of legal loading to cause the aircraft CG to move outside of either its forward or aft limits.

Adverse-Loaded CG Checks

Many modern aircraft have multiple rows of seats and often more than one baggage compartment. After any repair or alteration that changes the weight and balance, the A&P mechanic or repairman must ensure that no legal condition of loading can move the CG outside of its allowable limits. To determine this, adverse-loaded CG checks must be performed and the results noted in the weight and balance revision sheet.

Item	Weight (lbs)	Moment Indexes (lb-in/100)	New CG (inches from datum)
Airplane	1,876.0	+677.2	
Radio removed	−12.2	−1.93	
Power supply removed	−9.2	−8.74	
ELT removed	−3.2	−4.32	
Radio installed	+8.4	+1.23	
ELT installed	+1.7	+2.29	
Passenger seat installed	+21.0	+20.37	
Total	1,882.5	+686.1	+36.4

Figure 5-4. *Weight and moment index changes caused by a typical alteration.*

Airplane EW and EWCG	1,876.0 lbs at +36.14
Engine METO horsepower	230
CG range	(+40.9) to (+46.0) at 3,100 lbs (+33.0) to (+46.0) at 2,250 lbs or less Straight line variation between points given
Empty-weight CG range	None
Maximum weight	3,100 lbs takeoff/flight 2,950 landing
Datum to LEMAC	25.98
MAC	58.00
No. of seats	4 (2 front at +34.0) (2 rear at +74.0)
Fuel capacity	92 gal (88 gal usable) two 46 gal integral tanks in wings at +48.2 See NOTE 1 for data on unusable fuel.
Minimum fuel	(METO HP ÷ 2) 115 lbs at +48
Maximum baggage	160 lbs Area A (100 lbs at +97.0) Area B (60 lbs at +116.0)
Oil capacity	12 qt (−15) (6 qt usable) See NOTE 1 for data on undrainable oil.

NOTE 1: The certificated empty weight and corresponding center of gravity location must include unusable fuel of 30 lbs (+46) and undrainable oil of 0 lbs.

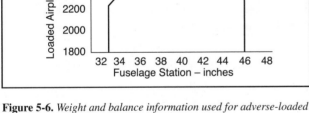

Figure 5-6. *Weight and balance information used for adverse-loaded CG Checks.*

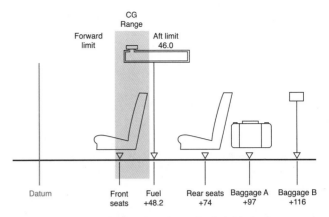

Figure 5-7. *Loading diagram for adverse-loaded CG check.*

Forward Adverse-Loaded CG Check

To conduct a forward CG check, make a chart that includes the airplane and any occupants and items of the load located in front of the forward CG limit. [Figure 5-7] Include only those items behind the forward limit that are essential to flight. This is the pilot and the minimum fuel.

In this example, the pilot, whose nominal weight is 170 pounds, is behind the forward CG limit. The fuel is also behind the forward limit, so the minimum fuel is used. For weight and balance purposes, the minimum fuel is no more than the quantity needed for one-half-hour of operation at rated maximum continuous power. This is considered to be 1/12 gallon for each maximum except takeoff (METO) horsepower. Because aviation gasoline weighs 6 pounds per gallon, determine the number of pounds of the minimum fuel by dividing the METO horsepower by 2; in this example minimum fuel is 115 pounds.

The front and rear seats and the baggage are all behind the forward CG limit, so no passengers or baggage are considered.

Make a chart like the one in Figure 5-8 to determine the CG with the aircraft loaded for its most forward CG. With the load consisting of only a pilot and the minimum fuel, the CG is +36.6, which is behind the most forward allowable limit for this weight of +33.0.

Item	Weight (lbs)	Arm (inches)	Moment (lb-in)	Most Forward CG +33.0
Airplane – empty	1,876.0	36.14	67,798.6	
Pilot	170.0	34.0	5,780.0	
Fuel (minimum)	115.0	48.0	5,520.0	
Total	2,161.0		79,098.6	+36.6

Figure 5-8. *Load conditions for forward adverse-loaded CG check.*

Aft Adverse-Loaded CG Check

To conduct an aft, or rearward, CG check, make a chart that includes the empty weight and EWCG of the aircraft after the alteration, and all occupants and items of the load behind the aft CG limit of 46.0. The pilot is in front of this limit, but is essential for flight and must be included. In this example, only the pilot will occupy the front seats. Since the CG of the fuel is behind the aft limit, full fuel will be used as well as the nominal weight (170 lbs) for both rear seat passengers and the maximum allowable baggage.

Under these loading conditions, the CG is located at +45.8, which is ahead of the aft limit of +46.0. [Figure5-9]

With only the pilot in front of the aft CG limit and maximum of all items behind the aft limit, the CG will be at +45.8 inches, which is ahead of the aft limit of +46.0 inches.

Ballast

It is possible to load most modern airplanes so the center of gravity shifts outside of the allowable limit. Placards and loading instructions in the Weight and Balance Data inform the pilot of the restrictions that will prevent such a shift from occurring. A typical placard in the baggage compartment of an airplane might read:

> When rear row of seats is occupied, 120 pounds of baggage or ballast must be carried in forward baggage compartment. For additional loading instructions, see Weight and Balance Data.

When the CG of an aircraft falls outside of the limits, it can usually be brought back in by using ballast.

Temporary Ballast

Temporary ballast, in the form of lead bars or heavy canvas bags of sand or lead shot, is often carried in the baggage compartments to adjust the balance for certain flight conditions. The bags are marked "Ballast XX Pounds - Removal Requires Weight and Balance Check." Temporary ballast must be secured so it cannot shift its location in flight, and the structural limits of the baggage compartment must not be exceeded. All temporary ballast must be removed before the aircraft is weighed.

Temporary Ballast Formula

The CG of a loaded airplane can be moved into its allowable range by shifting passengers or cargo, or by adding temporary ballast.

To determine the amount of temporary ballast needed, use this formula:

$$\text{Ballast weight needed} = \frac{\text{Total weight} \times \text{Distance needed to shift CG}}{\text{Distance between ballast location and desired CG}}$$

Permanent Ballast

If a repair or alteration causes the aircraft CG to fall outside of its limit, permanent ballast can be installed. Usually, permanent ballast is made of blocks of lead painted red and marked "Permanent Ballast - Do Not Remove." It should be attached to the structure so that it does not interfere with any control action, and attached rigidly enough that it cannot be dislodged by any flight maneuvers or rough landing.

Item	Weight (lbs)	Arm (inches)	Moment (lb-in)	Most Aft CG +46.0
Airplane – empty	1,876.0	36.14	67,798.6	
Pilot	170.0	34.0	5,780.0	
Fuel (full tanks – 88 gal.)	528.0	48.2	25,449.6	
Rear seat occupants (2)	340.0	74.0	25,160.0	
Baggage A	100.0	97.0	9,700.0	
Baggage B	60.0	116.0	6,960.0	
Total	3,074.0		140,848.2	+45.8

Figure 5-9. *Load conditions for aft adverse-loaded CG check.*

Two things must first be known to determine the amount of ballast needed to bring the CG within limits: the amount the CG is out of limits, and the distance between the location of the ballast and the limit that is affected.

If an airplane with an empty weight of 1,876 pounds has been altered so its EWCG is +32.2, and CG range for weights up to 2,250 pounds is +33.0 to +46.0, permanent ballast must be installed to move the EWCG from +32.2 to +33.0. There is a bulkhead at fuselage station 228 strong enough to support the ballast.

To determine the amount of ballast needed, use this formula:

$$\text{Ballast weight} = \frac{\text{Aircraft empty weight} \times \text{Dist. out of limits}}{\text{Distance between ballast and desired CG}}$$

$$= \frac{1,876 \times 0.8}{228 - 33}$$

$$= \frac{1,500.8}{195}$$

$$= 7.7 \text{ pounds}$$

A block of lead weighing 7.7 pounds attached to the bulkhead at fuselage station 228, will move the EWCG back to its proper forward limit of +33. This block should be painted red and marked "Permanent Ballast - Do Not Remove."

Weight and Balance Control – Helicopter

Weight and balance considerations of a helicopter are similar to those of an airplane, except they are far more critical, and the CG range is much more limited. [Figure 6-1] The engineers who design a helicopter determine the amount of cyclic control power that is available, and establish both the longitudinal and lateral CG envelopes that allow the pilot to load the helicopter so there is sufficient cyclic control for all flight conditions.

If the CG is ahead of the forward limit, the helicopter will tilt, and the rotor disk will have a forward pull. To counteract this, rearward cyclic is required. If the CG is too far forward, there may not be enough cyclic authority to allow the helicopter to flare for a landing, and it will consequently require an excessive landing distance.

If the CG is aft of the allowable limits, the helicopter will fly with a tail-low attitude and may need more forward cyclic stick displacement than is available to maintain a hover in a no-wind condition. There might not be enough cyclic power to prevent the tail boom striking the ground. If gusty winds should cause the helicopter to pitch up during high speed flight, there might not be enough forward cyclic control to lower the nose.

Helicopters are approved for a specific maximum gross weight, but it is not safe to operate them at this weight under all conditions. A high density altitude decrease the safe maximum weight as it affects the hovering, takeoff, climb, autorotation, and landing performance.

The fuel tanks on some helicopters are behind the CG, causing it to shift forward as fuel is used. Under some flight conditions, the balance may shift enough that there will not be sufficient cyclic authority to flare for landing. For these helicopters, the loaded CG should be computed for both takeoff and landing weights.

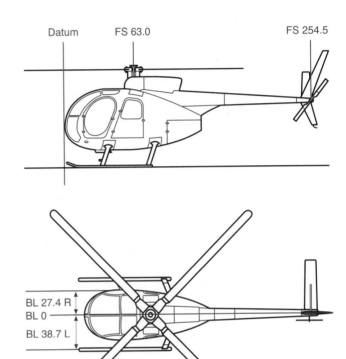

Figure 6-1. *Typical helicopter datum, flight stations, and butt line locations.*

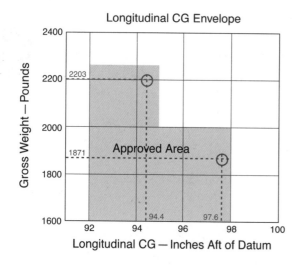

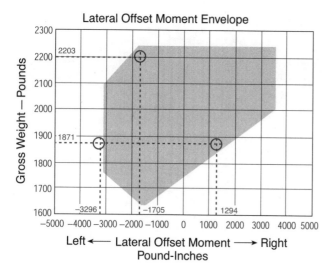

Figure 6-2. *Typical helicopter CG envelopes.*

Lateral balance of an airplane is usually of little concern and is not normally calculated. But some helicopters, especially those equipped for hoist operations, are sensitive

to the lateral position of the CG, and their POH/AFM include both longitudinal and lateral CG envelopes as well as information on the maximum permissible host load. Figure 6-2 is an example of such CG envelopes.

Determining the Loaded CG of a Helicopter

The empty weight and empty-weight center of gravity of a helicopter are determined in the same way as for an airplane. The weights recorded on the scales supporting the helicopter are added and their distance from the datum are used to compute the moments at each weighing point. The total moment is divided by the total weight to determine the location of the CG in inches from the datum. The datum of some helicopters is located at the center of the rotor mast, but since this causes some arms to be positive (behind the datum) and others negative (ahead of the datum), most modern helicopters have the datum located ahead of the aircraft, as do most modern airplanes. When the datum is ahead of the aircraft, all longitudinal arms are positive.

The lateral CG is determined in the same way as the longitudinal CG, except the distances between the scales and butt line zero (BL 0) are used as the arms. Arms to the right of BL 0 are positive and those to the left are negative. The Butt line zero (or sometimes referred to as the buttock) is a line through the symmetrical center of an aircraft from nose to tail. It serves as the datum for measuring the arms used to find the lateral CG. Lateral moments that cause the aircraft to rotate clockwise are positive (+), and those that cause it to rotate counter-clockwise are negative (-).

In order to determine whether or not a helicopter with the following specifications is within both longitudinal and lateral weight and balance limits, construct a chart like the one in Figure 6-3:

Item	Weight	Longitude Arm	Latitude Arm	Longitude Moment	Longitude CG	Lateral Offset Moment
Helicopter empty weight	1,545	101.4	+0.2	156,663		309
Pilot	170	64.0	−13.5	10,880		−2,295
Passenger	200	64.0	+13.5	12,800		2,700
Fuel 48 gallons	288	96.0	−8.4	27,648		−2,419
	2,203			207,991	94.4	−1,705

Figure 6-3. *Determining the longitudinal CG and the lateral offset moment.*

Empty weight.. 1,545 lbs
Empty -weight CG........... 101.4 in. aft of the datum
Lateral balance arm....................0.2 in. right of BL 0
Maximum allowable gross weight............. 2,250 lbs
Pilot............................170 lbs @64 in. aft of datum
    and 13.5 in. left of BL 0
Passenger ....200 lbs @ 64 in. aft of datum
    and 13.5 in. right of BL 0
Fuel 48 gal ....288 lbs @ 96 in. aft of datum
    and 84in. left of BL 0

Check the helicopter CG envelopes in Figure 6-2 to determine whether or not the CG is within limits both longitudinally and laterally.

In the longitudinal CG envelope, draw a line vertically upward from the CG of 94.4 inches aft of datum and a horizontal line from the weight of 2,203 pounds gross weight. These lines cross within the approved area.

In the lateral offset moment envelope, draw a line vertically upward from left, or -1,705 lb-in, and a line horizontally from 2,203 pounds on the gross weight index.

These lines cross within the envelope, showing the lateral balance is also within limits.

Effects of Offloading Passengers and Using Fuel

Consider the helicopter in Figure 6-3. The first leg of the flight consumes 22 gallons of fuel, and at the end of this leg, the passenger deplanes. Is the helicopter still within allowable CG limits for takeoff?

To find out, make a new chart like the one in Figure 6-4 to show the new loading conditions of the helicopter at the beginning of the second leg of the flight.

Under these conditions, according to the helicopter CG envelopes in Figure 6-2, the longitudinal CG is within limits. However, the lateral offset moment is excessive since both the pilot and the fuel are on the left side of the aircraft. If the POH allows it, the pilot may fly the aircraft on its second leg from the right-hand seat. According to Figures 6-5 and 6-2, this will bring the lateral balance into limits.

Item	Weight	Longitude Arm	Latitude Arm	Longitude Moment	Longitude CG	Lateral Offset Moment
Helicopter empty weight	1,545	101.4	+0.2	156,663		309
Pilot	170	64.0	−13.5	10,880		−2,295
Fuel 26 gallons	156	96.0	−8.4	14,976		−1,310
	1,871			182,519	97.6	−3,296

Figure 6-4. *Determining the longitudinal CG and the lateral offset moment for the second leg of the flight.*

Item	Weight	Longitude Arm	Latitude Arm	Longitude Moment	Longitude CG	Lateral Offset Moment
Helicopter empty weight	1,545	101.4	+0.2	156,663		309
Pilot	170	64.0	+13.5	10,880		2,295
Fuel 26 gallons	156	96.0	−8.4	14,976		−1,310
	1,871			182,519	97.6	1,294

Figure 6-5. *Determining the longitudinal CG and the lateral offset moment for the second leg of the flight with pilot flying from the right seat.*

Large Aircraft
Weight and Balance

The following consists of general guidelines for the weighing procedures of airplanes weighing over 12,500 pounds. Also included are several examples of center of gravity determination for various operational aspects of these aircraft. Persons seeking approval for a weight and balance control program for aircraft operated under Title 14 of the Code of Federal Regulations (14 CFR) part 91, subpart K, or parts 121, 125, and 135 should consult with the Flight Standards District Office (FSDO) or Certificate Management Office (CMO) having jurisdiction in their area.

Weighing Procedures

When weighing large aircraft, compliance with the relevant manuals, operations specifications, or management specification is required to ensure that weight and balance requirements specified in the aircraft flight manual (AFM) are met in accordance with approved limits. This will provide information to the flightcrew that allows the maximum payload to be carried safely.

The aircraft shall be weighed in an enclosed building after the aircraft has been cleaned. Check that the aircraft is in a configuration for weighing with regard to flight controls, unusable fuel, ballast, oil, and other operating fluids, and equipment as required by the controlling weight and balance procedure.

Large aircraft are not usually raised off the floor on jacks for weighing, they are weighed on ramp-type scales. The scales must be properly calibrated, zeroed, and used in accordance with the manufacturer's instructions. Each scale should be periodically checked for accuracy as recommended in the manufacturer's calibration schedule either by the manufacturer, or by a recognized facility such as a civil department of weights and measures. If no manufacturer's schedule is available, the period between calibrations should not exceed 1 year.

Determining the Empty Weight and EWCG

When the aircraft is properly prepared for weighing, roll it onto the scales, and level it. The weights are measured at three weighing points: the two main wheel points and the nose wheel point.

The empty weight and EWCG are determined by using the following steps, and the results are recorded in the weight and balance record for use in all future weight and balance computations.

1. Determine the moment index of each of the main-wheel points by multiplying the net weight (scale reading less tare weight), in pounds, at these points by the distance from the datum, in inches. Divide these numbers by the appropriate reduction factor.

2. Determine the moment index of the nose wheel weighing point by multiplying its net weight, in pounds, by its distance from the datum, in inches. Divide this by the reduction factor.

3. Determine the total weight by adding the net weight of the three weighing points and the total moment index by adding the moment indexes of each point.

4. Divide the total moment index by the total weight, and multiply this by the reduction factor. This gives the CG in inches, from the datum.

5. Determine the distance of the CG behind the leading edge of the mean aerodynamic chord (LEMAC) by subtracting the distance between the datum and LEMAC from the distance between the datum and the CG.

   Distance CG to LEMAC = Datum to CG − datum to LEMAC

6. Determine the EWCG in % MAC by using this formula:

   $$\text{EWCG in \% MAC} = \frac{\text{CG in inches from LEMAC} \times 100}{\text{MAC}}$$

Determining the Loaded CG of the Airplane in Percent MAC

The basic operating weight (BOW) and the operating index are entered into a loading schedule like the one in Figure 7-1 and the variables for the specific flight are entered as are appropriate to determine the loaded weight and CG.

Use the data in this example:

Basic operating Weight	105,500 lbs.
Basic operating index (total moment/1,000)	98,837.0
MAC	180.9 in
LEMAC	860.5

Item		Weight	Moment/1000
BOW		105,500	92,837
PAX forward	18	3,060	1,781
PAX aft	95	16,150	16,602
Fwd cargo		1,500	1,020
Aft cargo		2,500	2,915
Fuel tank 1		10,500	10,451
Fuel tank 3		10,500	10,451
Fuel tank 2		28,000	25,589
		177,710	161,646

Figure 7-1. *Loading tables.*

Use Figure 7-2 to determine the moment indexes for the passengers (PAX), cargo, and fuel.

The airplane is loaded in this way:

Passengers (nominal weight 170 pounds each)
Forward compartment 18
Aft compartment 95

Cargo
Forward hold .... 1,500 lbs
Aft hold .... 2,500 lbs

Fuel
Tank 1 & 3 .... 10,500 lbs each
Tank 2 .... 28,000 lbs

Determine the location of the CG in inches aft of the datum by using this formula:

$$\text{CG in. aft of datum} = \left(\frac{\text{Total moment index}}{\text{Total weight}}\right) \times 1,000$$

$$= \left(\frac{161,646}{177,710}\right) \times 1,000$$

$$= 909.6 \text{ inches}$$

Determine the distance from the CG to the LEMAC by subtracting the distance between the datum and LEMAC from the distance between the datum and the CG:

$$\text{Distance CG to LEMAC} = \text{Datum to CG} - \text{datum to LEMAC}$$

$$= 909.6 - 860.5$$

$$= 49.1 \text{ inches}$$

The location of the CG in percent of MAC must be known in order to set the stabilizer trim takeoff. Use this formula:

$$\text{CG \% MAC} = \left(\frac{\text{Distance CG to LEMAC}}{\text{MAC}}\right) \times 100$$

$$= \left(\frac{49.1}{180.9}\right) \times 100$$

$$= 27.1\%$$

On Board Aircraft Weighing System

Some large transport airplanes have an on board aircraft weighing system (OBAWS) that, when the aircraft is on the ground, gives the flightcrew a continuous indication of the aircraft total weight and the location of the CG in % MAC.

The system consists of strain-sensing transducers in each main wheel and nose wheel axle, a weight and balance computer, and indicators that show the gross weight, the CG location in percent of MAC, and an indicator of the ground attitude of the aircraft.

The strain sensors measure the amount each axle deflects and sends this data into the computer, where signals from all of the transducers and the ground attitude sensor are integrated. The results are displayed on the indicators for the flightcrew.

PASSENGER LOADING TABLE

Number of Pass.	Weight lbs	Moment 1000
Forward Compartment Centroid — 582.0		
5	850	495
10	1,700	989
15	2,550	1,484
20	3,400	1,979
25	4,250	2,473
29	4,930	2,869
AFT Compartment Centroid — 1028.0		
10	1,700	1,748
20	3,400	3,495
30	5,100	5,243
40	6,800	6,990
50	8,500	8,738
60	10,200	10,486
70	11,900	12,233
80	13,600	13,980
90	15,300	15,728
100	17,000	17,476
110	18,700	19,223
120	20,400	20,971
133	22,610	23,243

CARGO LOADING TABLE

	Moment 1000	
	Forward Hold	Aft Hold
Weight lbs	Arm 680.0	Arm 1166.0
6,000		6,966
5,000	3,400	5,830
4,000	2,720	4,664
3,000	2,040	3,498
2,000	1,360	2,332
1,000	680	1,166
900	612	1,049
800	544	933
700	476	816
600	408	700
500	340	583
400	272	466
300	204	350
200	136	233
100	68	117

FUEL LOADING TABLE

TANKS 1 & 3 (EACH)			TANKS 2 (3 CELL)					
Weight lbs	Arm	Moment 1000	Weight lbs	Arm	Moment 1000	Weight lbs	Arm	Moment 1000
8,500	992.1	8,433	8,500	917.5	7,799	22,500	914.5	20,576
9,000	993.0	8,937	9,000	917.2	8,255	23,000	914.5	21,034
9,500	993.9	9,442	9,500	917.0	8,711	23,500	914.4	21,488
10,000	994.7	9,947	10,000	916.8	9,168	24,000	914.3	21,943
10,500	995.4	10,451	10,500	916.6	9,624	24,500	914.3	22,400
11,000	996.1	10,957	11,000	916.5	10,082	25,000	914.2	22,855
11,500	996.8	11,463	11,500	916.3	10,537	25,500	914.2	23,312
12,000	997.5	11,970	12,000	916.1	10,993	26,000	914.1	23,767
						26,500	914.1	24,244
FULL CAPACITY			**(See note at lower left)			27,000	914.0	24,678
			18,500	915.1	16,929	27,500	913.9	25,132
**Note: Computations for Tank 2 weights for 12,500 lbs to 18,000 lbs have been purposely omitted.			19,000	915.0	17,385	28,000	913.9	25,589
			19,500	914.9	17,841	28,500	913.8	26,043
			20,000	914.9	18,298	29,000	913.7	26,497
			20,500	914.8	18,753	29,500	913.7	26,954
			21,000	914.7	19,209	30,000	913.6	27,408
			21,500	914.6	19,664			
			22,000	914.6	20,121	FULL CAPACITY		

Figure 7-2. *Loading schedule for determining weight and CG.*

Determining the Correct Stabilizer Trim Setting

It is important before takeoff to set the stabilizer trim for the existing CG location. There are two ways the stabilizer trim setting systems may be calibrated: in % MAC, and in Units ANU (Airplane Nose Up).

Stabilizer Trim Setting in Percent of MAC

If the stabilizer trim is calibrated in units of % MAC, determine the CG location in % MAC as has just been described, then set the stabilizer trim on the percentage figure thus determined.

Stabilizer Trim Setting in Percent of ANU (Airplane Nose Up)

Some aircraft give the stabilizer trim setting in Units ANU that correspond with the location of the CG in % MAC. When preparing for takeoff in an aircraft equipped with this system, first determine the CG in % MAC in the way described above, then refer to the Stabilizer Trim Setting Chart on the Takeoff Performance page of the AFM. Figure 7-3 is an excerpt from such a page from the AFM of a Boeing 737.

Consider an airplane with these specifications:

CG location......... station 635.7
LEMAC .... station 625
MAC .... 134.0 in

First determine the distance from the CG to the LEMAC by using this formula:

$$\text{Distance CG to LEMAC} = \text{Datum to CG} - \text{datum to LEMAC}$$
$$= 635.7 - 625.0$$
$$= 10.7 \text{ inches}$$

Then determine the location of the CG in percent of MAC by using this formula:

$$\text{CG \% MAC} = \left(\frac{\text{Distance CG to LEMAC}}{\text{MAC}}\right) \times 100$$
$$= \left(\frac{10.7}{134.0}\right) \times 100$$
$$= 8.0\% \text{ MAC}$$

Refer to Figure 7-3. For all flap settings and a CG located at 8% MAC, the stabilizer setting is 7¾ Units ANU.

Stabilizer Trim Setting — Units Airplane Nose Up	
CG	Flaps (All)
6	8
8	7¾
10	7½
12	7
14	6¾
16	6¼
18	5¾
20	5½
22	5
24	4½
26	4
28	3½
30	3
32	2½

Figure 7-3. *Stabilizer trim setting in ANU units.*

Determining CG Changes Caused by Modifying the Cargo

Large aircraft carry so much cargo that adding, subtracting, or moving any of it from one hold to another can cause large shifts in the CG.

Effects of Loading or Offloading Cargo

Both the weight and CG of an aircraft are changed when cargo is offloaded or onloaded. This example shows the way to determine the new weight and CG after 2,500 pounds of cargo is offloaded from the forward cargo hold.

Consider these specifications:

Loaded weight .... 90,000 lbs
Loaded CG................................ 22.5% MAC
Weight change................................- 2,500 lbs
Fwd. cargo hold centroid .... station 352.1
MAC ....141.1 in
LEMAC .... station 549.13

1. Determine the CG location in inches from the datum before the cargo is removed. Do this by first determining the distance of the CG aft of the LEMAC:

$$\text{CG (in. aft of LEMAC)} = \left(\frac{\text{CG in \% MAC}}{100}\right) \times \text{MAC}$$
$$= \left(\frac{22.5}{100}\right) \times 141.5$$
$$= 31.84 \text{ inches}$$

2. Determine the distance between the CG and the datum by adding the CG in inches aft of LEMAC to the distance from the datum to LEMAC:

CG (in. from datum) = CG in. aft of LEMAC +
datum to LEMAC

= 31.84 + 549.13

= 580.97 inches

3. Determine the moment/1,000 for the original weight:

$$\text{Moment}/1{,}000 = \frac{\text{Weight} \times \text{Arm}}{1{,}000}$$

$$= \frac{90{,}000 \times 580.97}{1{,}000}$$

$$= 52{,}287.30$$

4. Determine the new weight and new CG by first determining the moment/1,000 of the removed weight.

Multiply the amount of weight removed (-2,500 pounds) by the centroid of the forward cargo hold (352.1 inches), and then divide this by 1,000.

$$\text{Moment}/1{,}000 = \frac{\text{Weight} \times \text{Arm}}{1{,}000}$$

$$= \frac{-2{,}500 \times 352.1}{1{,}000}$$

$$= -880.25$$

5. Subtract the removed weight and its moment/1,000 from the original weight and moment/1,000.

Item	Weight	Moment/1000
Original weight	90,000	52,287.30
Δ Weight	− 2,500	− 880.25
New weight & moment	87,500	51,407.05

6. Determine the location of the new CG by dividing the total moment/1,000 by the total weight and multiplying this by the reduction factor 1,000.

$$CG = \left(\frac{\text{Total moment}/1{,}000}{\text{Total weight}}\right) \times 1{,}000$$

$$= \left(\frac{51{,}407}{87{,}500}\right) \times 1{,}000$$

$$= 587.5 \text{ inches behind the datum}$$

7. Convert the new CG location to % MAC. First, determine the distance between the CG location and LEMAC:

CG (in. aft of LEMAC) = CG (in. from datum) −
LEMAC

= 587.5 − 549.13

= 38.37 inches

8. Then, determine new CG in % MAC:

$$CG \% \text{ MAC} = \left(\frac{\text{Distance CG to LEMAC}}{\text{MAC}}\right) \times 100$$

$$= \left(\frac{38.37}{141.5}\right) \times 100$$

$$= 27.1\% \text{ MAC}$$

Offloading 2,500 pounds of cargo from the forward cargo hold moves the CG from 22.5% MAC to 27.1% MAC.

Effects of Onloading Cargo

The previous example showed the way the weight and CG changed when cargo was offloaded. This example shows the way both parameters change when cargo is onloaded.

The same basic airplane is used in the following example, but 3,000 pounds of cargo is onloaded in the forward cargo hold.

Weight before cargo is loaded....... 87,500 lbs
CG before cargo is loaded .... 27.1% MAC
Weight change.............................+ 3,000 lbs
Fwd. cargo hold centroid .... station 352.1
MAC ....141.5 in
LEMAC .... station 549.13

1. Determine the CG location in inches from the datum before the cargo is onloaded. Do this by first determining the distance of the CG aft of the LEMAC:

$$\begin{aligned}CG \text{ (inches aft} \atop \text{of LEMAC)} &= \left(\frac{CG \text{ in } \% \text{ MAC}}{100}\right) \times \text{MAC}\end{aligned}$$

$$= \left(\frac{27.1}{100}\right) \times 141.5$$

$$= 38.35 \text{ inches}$$

2. Determine the distance between the CG and the datum by adding the CG in inches aft of LEMAC to the distance from the datum to LEMAC:

CG (in. from datum) = CG in. aft of LEMAC +
datum to LEMAC

= 38.35 + 549.13

= 587.48 inches

3. Determine the moment/1,000 for the original weight:

$$\text{Moment}/1{,}000 = \frac{\text{Weight} \times \text{Arm}}{1{,}000}$$

$$= \frac{87{,}500 \times 587.48}{1{,}000}$$

$$= 51{,}404.5$$

4. Determine the new weight and new CG by first determining the moment/1,000 of the added weight. Multiply the amount of weight added (3,000 pounds) by the centroid of the forward cargo hold (352.1 inches), and then divide this by 1,000.

$$\text{Moment}/1,000 = \frac{\text{Weight} \times \text{Arm}}{1,000}$$

$$= \frac{3,000 \times 352.1}{1,000}$$

$$= 1,056.3$$

5. Add the onloaded cargo weight and its moment/1,000 to the original weight and moment/1,000.

	Weight	Moment/1000	CG in/datum	CG % MAC
Original weight and CG	87,500	51,404.5	587.48	27.1
Δ Weight	+ 3,000	1,056.3		
New weight and CG	90,500	52,460.8	579.68	21.59

6. Determine the location of the new CG by dividing the total moment/1,000 by the total weight and multiplying this by the reduction factor of 1,000.

$$CG = \frac{\text{Total moment}/1,000}{\text{Total weight}} \times 1,000$$

$$= \frac{52,460.8}{90,500} \times 1,000$$

$$= 579.68 \text{ inches behind the datum}$$

7. Convert the new CG location to % MAC. First, determine the distance between the CG location and LEMAC:

$$CG \text{ (in. aft of LEMAC)} = CG \text{ (in. from datum)} - LEMAC$$

$$= 579.68 - 549.13$$

$$= 30.55 \text{ inches}$$

8. Then, determine new CG in % MAC:

$$CG \% MAC = \left(\frac{\text{Distance CG to LEMA}}{\text{MAC}}\right) \times 100$$

$$= \left(\frac{30.55}{141.5}\right) \times 100$$

$$= 21.59\% \text{ MAC}$$

Onloading 3,000 pounds of cargo into the forward cargo hold moves the CG forward 5.51 inches, from 27.1% MAC to 21.59% MAC.

Effects of Shifting Cargo from One Hold to Another

When cargo is shifted from one cargo hold to another, the CG changes, but the total weight of the aircraft remains the same.

As an example, use this data:

Loaded weight........................... 90,000 lbs
Loaded CG .... station 580.97
  (which is 22.5% MAC)
Fwd. cargo hold centroid .... station 352
Aft cargo hold centroid .... station 724.9
MAC...141.5 in
LEMAC.................................... station 549

To determine the change in CG, or __CG, caused by shifting 2,500 pounds of cargo from the forward cargo hold to the aft cargo hold, use this formula:

$$\Delta CG = \frac{\text{Weight shifted} \times \text{Distance shifted}}{\text{Total weight}}$$

$$= \frac{2,500 \times (227.9 + 144.9)}{90,000}$$

$$= \frac{2,500 \times 372.8}{90,000}$$

$$= 10.36 \text{ inches}$$

Since the weight was shifted aft, the CG moved aft, and the CG change is positive. If the shift were forward, the CG change would be negative.

Before the cargo was shifted, the CG was located at station 580.97, which is 22.5% MAC. The CG moved aft 10.36 inches, so the new CG is:

$$\text{New CG} = \text{Old CG} \pm \Delta CG$$

$$= 580.97 + 10.36$$

$$= 591.33 \text{ inches}$$

Convert the location of the CG in inches aft of the datum to % MAC by using this formula:

$$\Delta CG\ \%\ MAC = \left(\frac{\Delta CG\ inches}{MAC}\right) \times 100$$
$$= \left(\frac{10.36}{141.5}\right) \times 100$$
$$= 7.32\%\ MAC$$

The new CG in % MAC caused by shifting the cargo is the sum of the old CG plus the change in CG:

$$New\ CG\ \%\ MAC = Old\ CG \pm \Delta CG$$
$$= 22.5\% + 7.32\%$$
$$= 29.82\%\ MAC$$

Some aircraft AFMs locate the CG relative to an index point rather than the datum or the MAC. An index point is a location specified by the aircraft manufacturer from which arms used in weight and balance computations are measured. Arms measured from the index point are called index arms, and objects ahead of the index point have negative index arms, while those behind the index point have positive index arms.

Use the same data as in the previous example, except for these changes:

Loaded CG .... index arm of 0.97, which is 22.5% MAC
Index point .... fuselage station 580.0
Fwd. cargo hold centroid .... -227.9 index arm
Aft cargo hold centroid .... +144.9 index arm
MAC .... 141.5 in
LEMAC .... -30.87 index arm

The weight was shifted 372.8 inches (-227.9 to +144.9 = 372.8).

The change in CG can be calculated by using this formula:

$$\Delta CG = \frac{Weight\ shifted \times Distance\ shifted}{Total\ weight}$$
$$= \frac{2,500 \times (724.9 - 352)}{90,000}$$
$$= \frac{2,500 \times 372.9}{90,000}$$
$$= 10.36\ inches$$

Since the weight was shifted aft, the CG moved aft, and the CG change is positive. If the shift were forward, the CG change would be negative.

Before the cargo was shifted, the CG was located at 0.97 index arm, which is 22.5% MAC. The CG moved aft 10.36 inches, and the new CG is:

$$New\ CG = Old\ CG \pm \Delta CG$$
$$= 0.97 + 10.36$$
$$= 11.33\ index\ arm$$

The change in the CG in % MAC is determined by using this formula:

$$New\ CG\ \%\ MAC = Old\ CG \pm \Delta CG$$
$$= 22.5\% + 7.32\%$$
$$= 29.82\%\ MAC$$

The new CG in % MAC is the sum of the old CG plus the change in CG:

$$\Delta CG\ \%\ MAC = \left(\frac{\Delta CG\ inches}{MAC}\right) \times 100$$
$$= \left(\frac{10.36}{141.5}\right) \times 100$$
$$= 7.32\%\ MAC$$

Notice that the new CG is in the same location whether the distances are measured from the datum or from the index point.

Determining Cargo Pallet Loads with Regard to Floor Loading Limits

Each cargo hold has a structural floor loading limit based on the weight of the load and the area over which this weight is distributed. To determine the maximum weight of a loaded cargo pallet that can be carried in a cargo hold, divide its total weight, which includes the weight of the empty pallet and its tiedown devices, by its area in square feet. This load per square foot must be equal to or less than the floor load limit.

In this example, determine the maximum load that can be placed on this pallet without exceeding the floor load limit.

Pallet dimensions .... 36 by 48 in
Empty pallet weight .... 47 lbs
Tiedown devices .... 33 lbs
Floor load limit .... 169 pounds per square foot

The pallet has an area of 36 inches (3 feet) by 48 inches (4 feet). This is equal to 12 square feet. The floor has a load limit of 169 pounds per square foot; therefore, the total weight of the loaded pallet can be 169 x 12 = 2,028 pounds.

Subtracting the weight of the pallet and the tiedown devices gives an allowable load of 1,948 pounds (2,028 - [47 + 33]).

Determine the floor load limit that is needed to carry a loaded cargo pallet having these dimensions and weights:

Pallet dimensions .... 48.5 by 33.5 in

Pallet weight .... 44 lbs

Tiedown devices .... 27 lbs

Cargo weight .... 786.5 lbs

First determine the number of square feet of pallet area:

$$\text{Area (sq. ft.)} = \frac{\text{Length (inches)} \times \text{Width (inches)}}{144}$$

$$= \frac{48.5 \times 33.5}{144}$$

$$= \frac{1,624.7}{144}$$

$$= 11.28 \text{ square feet}$$

Then determine the total weight of the loaded pallet:

Pallet	44.0 lbs
Tiedown devices	27.0 lbs
Cargo	786.5 lbs
	857.5 lbs

Determine the load imposed on the floor by the loaded pallet:

The floor must have a minimum load limit of 76 pounds per square foot.

$$\text{Floor Load} = \frac{\text{Loaded weight}}{\text{Pallet area}}$$

$$= \frac{857.5}{11.28}$$

$$= 76.0 \text{ pounds/square foot}$$

Determining the Maximum Amount of Payload That Can Be Carried

The primary function of a transport or cargo aircraft is to carry payload. This is the portion of the useful load, passengers or cargo, that produces revenue. To determine the maximum amount of payload that can be carried, follow a series of steps, considering both the maximum limits for the aircraft and the trip limits imposed by the particular trip. In each step, the trip limit must be less than the maximum limit. If it is not, the maximum limit must be used.

These are the specifications for the aircraft in this example:

Basic operating weight (BOW) .... 100,500 lbs

Maximum zero fuel weight .... 138,000 lbs

Maximum landing weight .... 142,000 lbs

Maximum takeoff weight .... 184,200 lbs

Fuel tank load .... 54,000 lbs

Est. fuel burn en route .... 40,000 lbs

1. Compute the maximum takeoff weight for this trip. This is the maximum landing weight plus the trip fuel.

Max. Limit		Trip Limit
142,000	Landing weight	142,000
	+ trip fuel	+ 40,000
184,200	Takeoff weight	182,000

2. The trip limit is the lower, so it is used to determine the zero fuel weight.

Max. Limit		Trip Limit
184,200	Takeoff weight	182,000
	- fuel load	-54,000
138,000	Zero fuel weight	128,000

3. The trip limit is again lower, so use it to compute the maximum payload for this trip.

Max. Limit		Trip Limit
138,000	Zero fuel weight	128,000
	- BOW	- 100,500
	Payload (pounds)	27,500

Under these conditions 27,500 pounds of payload may be carried.

Determining the Landing Weight

It is important to know the landing weight of the airplane in order to set up the landing parameters, and to be certain the airplane will be able to land at the intended destination.

In this example of a four-engine turboprop airplane, determine the airplane weight at the end of 4.0 hours of cruise under these conditions:

Takeoff weight .... 140,000 lbs

Pressure altitude during cruise .... 16,000 feet

Ambient temperature during cruise .... -32°C

Fuel burned during descent and landing ..... 1,350 lbs

Determine the weight at the end of cruise by using the Gross Weight Table of Figure 7-4 and following these steps:

1. Use the U.S. Standard Atmosphere Table in Figure 7-5 to determine the standard temperature for 16,000 feet. This is -16.7°C.

2. The ambient temperature is -32°C, which is a deviation from standard of 15.3°C. (-32° – -16.7° = 15.3°). It is below standard.

3. In Figure 7-4, follow the vertical line representing 140,000 pounds gross weight upward until it intersects the diagonal line for 16,000 feet pressure altitude.

4. From this intersection, draw a horizontal line to the left to the temperature deviation index (0°C deviation).

5. Draw a diagonal line parallel to the dashed lines for "Below Standard" from the intersection of the horizontal line and the Temperature Deviation Index.

6. Draw a vertical line upward from the 15.3°C Temperature Deviation From Standard.

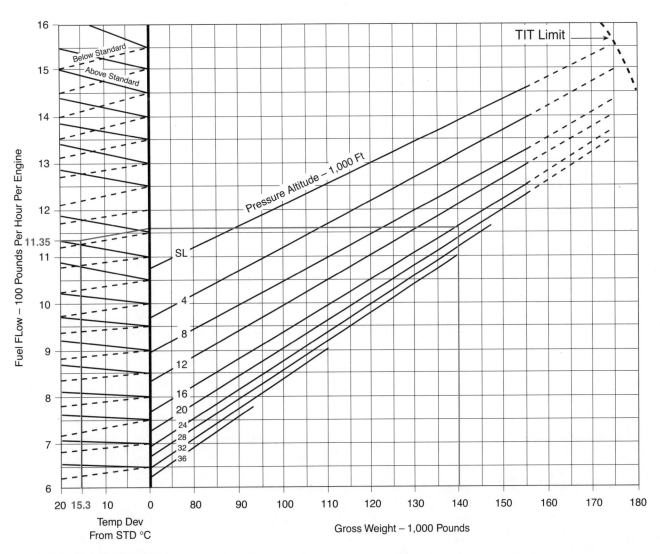

Figure 7-4. *Gross Weight Table.*

7. Draw a horizontal line to the left from the intersection of the "Below Standard" diagonal and the 15.3°C temperature deviation vertical line. This line crosses the "Fuel Flow-100 Pounds per Hour per Engine" index at 11.35. This indicates that each of the four engines burns 1,135 (100 x 11.35) pounds of fuel per hour. The total fuel burn for the 4-hour cruise is:

Total fuel burn = Lb/hr/engine × No. engines ×
Hours flight duration
= 1,135 × 4 × 4
= 18,160 pounds

8. The airplane gross weight was 140,000 pounds at takeoff, and since 18,160 pounds of fuel was burned during cruise and 1,350 pounds was burned during the approach and landing phase, the landing weight is:

140,000 - (18,160 + 1,350) = 120,490 pounds

TABLE OF U.S. STANDARD ATMOSPHERE

Feet	inHg	mmHg	PSI	°C	°F
0	29.92	760.0	14.70	15.0	59.0
2,000	27.82	706.7	13.66	11.0	51.9
4,000	25.84	656.3	12.69	7.1	44.7
6,000	23.98	609.1	11.78	3.1	37.6
8,000	22.23	564.6	10.92	-0.8	30.5
10,000	20.58	522.7	10.11	-4.8	23.3
12,000	19.03	483.4	9.35	-8.8	16.2
14,000	17.58	446.5	8.63	-12.7	9.1
16,000	16.22	412.0	7.96	-16.7	1.9
18,000	14.95	379.7	7.34	-20.7	-5.2
20,000	13.76	349.5	6.75	-24.6	-12.3
22,000	12.65	321.3	6.21	-28.6	-19.5
24,000	11.61	294.9	5.70	-32.5	-26.6
26,000	10.64	270.3	5.22	-36.5	-33.7
28,000	9.74	237.4	4.78	-40.5	-40.9
30,000	8.90	226.1	4.37	-44.4	-48.0
32,000	8.12	206.3	3.98	-48.4	-55.1
34,000	7.40	188.0	3.63	-52.4	-62.3
36,000	6.73	171.0	3.30	-56.3	-69.4
38,000	6.12	155.5	2.99	-56.5	-69.7
40,000	5.56	141.2	2.72	-56.5	-69.7
42,000	5.05	128.3	2.47	-56.5	-69.7
44,000	4.59	116.6	2.24	-56.5	-69.7
46,000	4.17	105.9	2.04	-56.5	-69.7
48,000	3.79	96.3	1.85	-56.5	-69.7
50,000	3.44	87.4	1.68	-56.5	-69.7
55,000	2.71	68.6	1.32	TEMPERATURE REMAINS CONSTANT	
60,000	2.14	54.4	1.04		

inHg = Inches of Mercury   °C = Centigrade

mmHg = Millimeter of Mercury   °F = Fahrenheit

PSI = Pounds per square inch

Figure 7-5. *Standard atmosphere table.*

Determining the Minutes of Fuel Dump Time

Most large aircraft are approved for a greater weight for takeoff than for landing, and to make it possible for them to return to landing soon after takeoff, a fuel jettison system is sometimes installed.

It is important in an emergency situation that the flightcrew be able to dump enough fuel to lower the weight to its allowed landing weight. This is done by timing the dumping process.

In this example, the aircraft has three engines operating and these specifications apply:

Cruise weight ....171,000 lbs

Maximum landing weight ....142,500 lbs

Time from start of dump to landing ....19 minutes

Average fuel flow during
dumping and descent.........................3,170 lb/hr/eng

Fuel dump rate ....2,300 pounds per minute

Follow these steps to determine the number of minutes of fuel dump time:

1. Determine the amount the weight of the aircraft must be reduced to reach the maximum allowable landing weight:

$$
\begin{array}{ll}
171,000 & \text{lbs cruise weight} \\
- \ 142,500 & \text{lbs maximum landing weight} \\
\hline
28,500 & \text{lbs required reduction}
\end{array}
$$

2. Determine the amount of fuel burned from the beginning of the dump to touchdown:

$$\text{Fuel flow} = \frac{3,170 \text{ lb/hr/engine}}{60}$$

$$= 52.83 \text{ lb/min engine}$$

For all three engines, this is 52.83° 3 = 158.5 lbs/min.

The three engines will burn 158.5° 19 = 3,011.5 pounds of fuel between the beginning of dumping and touchdown.

3. Determine the amount of fuel needed to dump by subtracting the amount of fuel burned during the dumping from the required weight reduction:

$$
\begin{array}{ll}
28,500.0 & \text{lbs required weight reduction} \\
- \ 3,011.5 & \text{lbs fuel burned after start of dumping} \\
\hline
25,488.5 & \text{lbs fuel to be dumped}
\end{array}
$$

4. Determine the time needed to dump this amount of fuel by dividing the number of pounds of fuel to dump by the dump rate:

$$\frac{25,488.5 \text{ lbs}}{2,300 \text{ lb/min}} = 11.08 \text{ minutes}$$

Weight and Balance of Commuter Category Airplanes

The Beech 1900 is a typical commuter category airplane that can be configured to carry passengers or cargo. Figure 7-6 shows the loading data of this type of airplane in the passenger configuration, and Figure 7-14 on Page 7-16 shows the cargo configuration. Jet fuel weight is affected by temperature, the colder the fuel, the more dense and therefore the more pounds of fuel per gallon. [Text continued on page 7-15.]

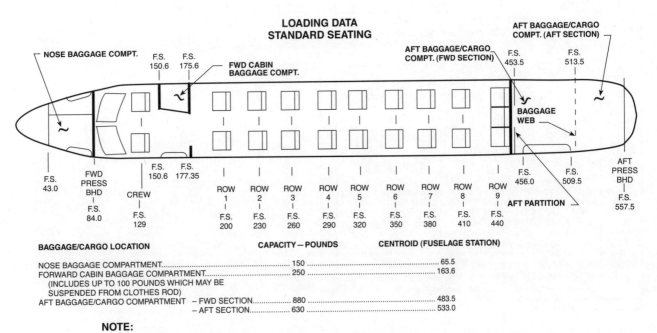

Figure 7-6. *Loading data for passenger configuration.*

Item	Weight	Arm	Moment/100	C G
Airplane basic EW	9,226		25,823	
Crew	340	129	439	
Passengers				
Row 1	300	200	600	
Row 2	250	230	575	
Row 3	190	260	494	
Row 4	170	290	493	
Row 5	190	320	608	
Row 6	340	350	1,190	
Row 7	190	380	722	
Row 8		410		
Row 9		440		
Baggage				
Nose		65.5		
Fwd Cabin	100	163.6	164	
Aft (Fwd Section)	200	483.5	967	
Aft (Aft Section)	600	533.0	3,198	
Fuel Jet A @ +25°C				
Gallons 390	2,633		7,866	
	14,729		43,139	292.9

Figure 7-7. *Determining the loaded weight and CG of a Beech 1900 in the passenger configuration.*

USEFUL LOAD WEIGHTS AND MOMENTS
OCCUPANTS

WEIGHT	CREW	CABIN SEATS								
	F.S. 129	F.S. 200	F.S. 230	F.S. 260	F.S. 290	F.S. 320	F.S. 350	F.S. 380	F.S. 410	F.S. 440
	MOMENT/100									
80	103	160	184	208	232	256	280	304	328	352
90	116	180	207	234	261	288	315	342	369	396
100	129	200	230	260	290	320	350	380	410	440
110	142	220	253	286	319	352	385	418	451	484
120	155	240	276	312	348	384	420	456	492	528
130	168	260	299	338	377	416	455	494	533	572
140	181	280	322	364	406	448	490	532	574	616
150	194	300	345	390	435	480	525	570	615	660
160	206	320	368	416	464	512	560	608	656	704
170	219	340	391	442	493	544	595	646	697	748
180	232	360	414	468	522	576	630	684	738	792
190	245	380	437	494	551	608	665	722	779	836
200	258	400	460	520	680	640	700	760	820	880
210	271	420	483	546	609	672	735	798	861	924
220	284	440	506	572	638	704	770	836	902	968
230	297	460	529	598	667	736	805	874	943	1012
240	310	480	552	624	696	768	840	912	984	1056
250	323	500	575	650	725	800	875	950	1025	1100

Note: Weights reflected in above table represent weight per seat.

Figure 7-8. *Weights and moments - occupants.*

USEFUL LOAD WEIGHTS AND MOMENTS
BAGGAGE

WEIGHT	NOSE BAGGAGE COMPART-MENT F.S. 65.5	FORWARD CABIN BAGGAGE COMPART-MENT F.S. 163.6	AFT BAGGAGE/CARGO COMPART-MENT (FORWARD SECTION) F.S. 483.5	AFT BAGGAGE/CARGO COMPART-MENT (AFT SECTION) F.S. 533.0
	MOMENT/100			
10	7	16	48	53
20	13	33	97	107
30	20	49	145	160
40	26	65	193	213
50	33	82	242	266
60	39	98	290	320
70	46	115	338	373
80	52	131	387	426
90	59	147	435	480
100	66	164	484	533
150	98	245	725	800
200		327	967	1066
250		409	1209	1332
300			1450	1599
350			1692	1866
400			1934	2132
450			2176	2398
500			2418	2665
550			2659	2932
600			2901	3198
630			3046	3358
650			3143	
700			3384	
750			3626	
800			3868	
850			4110	
880			4255	

Figure 7-9. *Weights and moments - baggage.*

DENSITY VARIATION OF AVIATION FUEL
BASED ON AVERAGE SPECIFIC GRAVITY

FUEL	AVERAGE SPECIFIC GRAVITY AT 15°C (59°F)
AVIATION KEROSENE JET A AND JET A1	.812
JET B (JP-4)	.785
AV GAS GRADE 100/130	.703

NOTE: The Fuel Quantity Indicator is calibrated for correct indication when using Aviation Kerosene Jet A and Jet A1. When using other fuels, multiply the indicated fuel quantity in pounds by .99 for Jet B (JP-4) or by .98 for Aviation Gasoline (100/130) to obtain actual fuel quantity in pounds.

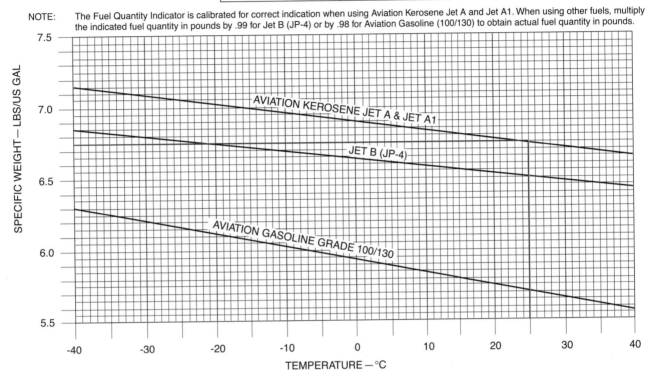

Figure 7-10. *Density variation of aviation fuel.*

USEFUL LOAD WEIGHTS AND MOMENTS
USABLE FUEL

GALLONS	6.5 LB/GAL WEIGHT	6.5 LB/GAL MOMENT 100	6.6 LB/GAL WEIGHT	6.6 LB/GAL MOMENT 100	6.7 LB/GAL WEIGHT	6.7 LB/GAL MOMENT 100	6.8 LB/GAL WEIGHT	6.8 LB/GAL MOMENT 100
10	65	197	66	200	67	203	68	206
20	130	394	132	401	134	407	136	413
30	195	592	198	601	201	610	204	619
40	260	789	264	802	268	814	272	826
50	325	987	330	1002	335	1018	340	1033
60	390	1185	396	1203	402	1222	408	1240
70	455	1383	462	1404	469	1426	476	1447
80	520	1581	528	1605	536	1630	544	1654
90	585	1779	594	1806	603	1834	612	1861
100	650	1977	660	2007	670	2038	680	2068
110	715	2175	726	2208	737	2242	748	2275
120	780	2372	792	2409	804	2445	816	2482
130	845	2569	858	2608	871	2648	884	2687
140	910	2765	924	2808	938	2850	952	2893
150	975	2962	990	3007	1005	3053	1020	3099
160	1040	3157	1056	3205	1072	3254	1088	3303
170	1105	3351	1122	3403	1139	3454	1156	3506
180	1170	3545	1188	3600	1206	3654	1224	3709
190	1235	3739	1254	3797	1273	3854	1292	3912
200	1300	3932	1320	3992	1340	4053	1360	4113
210	1365	4124	1386	4187	1407	4250	1428	4314
220	1430	4315	1452	4382	1474	4448	1496	4514
230	1495	4507	1518	4576	1541	4646	1564	4715
240	1560	4698	1584	4770	1608	4843	1632	4915
250	1625	4889	1650	4964	1675	5040	1700	5115
260	1690	5080	1716	5158	1742	5236	1768	5315
270	1755	5271	1782	5352	1809	5433	1836	5514
280	1820	5462	1848	5546	1876	5630	1904	5714
290	1885	5651	1914	5738	1943	5825	1972	5912
300	1950	5842	1980	5932	2010	6022	2040	6112
310	2015	6032	2046	6125	2077	6218	2108	6311
320	2080	6225	2112	6321	2144	6416	2176	6512
330	2145	6417	2178	6516	2211	6615	2244	6713
340	2210	6610	2244	6711	2278	6813	2312	6915
350	2275	6802	2310	6907	2345	7011	2380	7116
360	2340	6995	2376	7103	2412	7210	2448	7318
370	2405	7188	2442	7299	2479	7409	2516	7520
380	2470	7381	2508	7495	2546	7609	2584	7722
390	2535	7575	2574	7691	2613	7808	2652	7924
400	2600	7768	2640	7888	2680	8007	2720	8127
410	2665	7962	2706	8085	2747	8207	2788	8330
420	2730	8156	2772	8282	2814	8407	2856	8532
425	2763	8259	2805	8386	2848	8513	2890	8640

Figure 7-11. *Weights and moments -usable fuel.*

WEIGHT AND BALANCE DIAGRAM

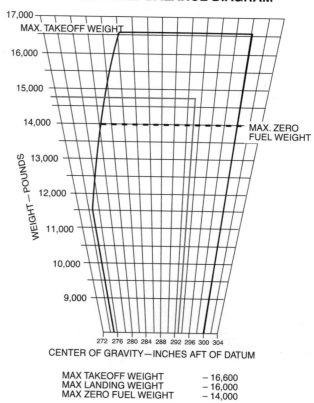

CENTER OF GRAVITY—INCHES AFT OF DATUM

MAX TAKEOFF WEIGHT — 16,600
MAX LANDING WEIGHT — 16,000
MAX ZERO FUEL WEIGHT — 14,000

Figure 7-12. *Weight and balance diagram.*

Item	Weight	Arm	Moment/100	CG
Row 1	(-) 300	200	(-) 600	
Row 2	(-) 250	230	(-) 575	
Row 8	(+) 300	410	(+) 1,230	
Row 9	(+) 250	440	(+) 1,100	
Original conditions	14,729		43,139	
Changes	0		(+) 1,155	
New conditions	14,729		44,294	300.7

Figure 7-13. *Change in CG caused by shifting passenger seats.*

Determining the Loaded Weight and CG

As this airplane is prepared for flight, a manifest like the one in Figure 7-7 is prepared.

1. The crew weight and the weight of each passenger is entered into the manifest, and the moment/100 for each occupant is determined by multiplying the weight by the arm and dividing by 100. This data is available in the AFM and is shown in the Weight and Moments- Occupants table in Figure 7-8 on Page 7-12.

2. The weight of the baggage in each compartment that is used is entered with its moment/100. This is determined in the Weights and Moments- Baggage table in Figure 7-9 on Page 7-12.

3. Determine the weight of the fuel. Jet A fuel has a nominal specific gravity at +15°C of 0.812 and weighs 6.8 pounds per gallon, but at +25°C, according to the chart in Figure 7-10 on Page 7-13, it weighs 6.75 lbs/ gal.

    Using Figure 7-11 on Page 7-14, determine the weights and moment/100 for 390 gallons of Jet A fuel by interpolating between those for 6.7 lbs/gal and 6.8 lbs/ gal. The 390 gallons of fuel at this temperature weighs 2,633 pounds, and its moment index is 7,866 lb-in/100.

4. Add all of the weights and all of the moment indexes. Divide the total moment index by the total weight, and multiply this by the reduction factor of 100. The total weight is 14,729 pounds, the total moment index is 43,139 lb-in/100. The CG is located at fuselage station 292.9.

5. Check to determine that the CG is within limits for this weight. Refer to the Weight and Balance Diagram in Figure 7-12 on Page 7-14. Draw a horizontal line across the envelope at 14,729 pounds of weight and a vertical line from the CG of 292.9 inches aft of datum. These lines cross inside the envelope verifying the CG is within limits for this weight.

Determining the Changes in CG When Passengers are Shifted

Consider the airplane above for which the loaded weight and CG have just been determined, and determine the change in CG when the passengers in rows 1 and 2 are moved to rows 8 and 9. Figure 7-13 shows the changes from the conditions shown in Figure 7-7. There is no weight change, but the moment index has been increased by 1,155 pound-inches/100 to 44,294. The new CG is at fuselage station 300.7.

$$CG = \left(\frac{43,139 + 1,155}{14,729} \right) \times 100$$
$$= 300.7$$

This type of problem is usually solved by using the following two formulas. The total amount of weight shifted is 550 pounds(300 + 250) and both rows of passengers have moved aft by 210 inches (410 - 200 and 440 - 230).

$$\Delta CG = \frac{\text{Weight shifted} \times \text{Distance shifted}}{\text{Total weight}}$$
$$= \frac{550 \times 210}{14,748}$$
$$= 7.8 \text{ inches}$$

$$CG = \text{Original CG} + \Delta CG$$
$$= 292.9 + 7.8$$
$$= 300.7 \text{ inches aft of datum}$$

The CG has been shifted aft 7.8 inches and the new CG is at station 300.7.

Determining Changes in Weight and CG When the Airplane is Operated in its Cargo Configuration

Consider the airplane configuration shown in Figure 7-14.

The airplane is loaded as recorded in the table in Figure 7-15. The basic operating weight (BOW) includes the pilots and their baggage so there is no separate item for them.

The arm of each cargo section is the centroid of that section, as shown in Figure 7-14.

The fuel, at the standard temperature of 15°C weighs 6.8 pounds per gallon. Refer to the Weights and Moments - Usable Fuel in Figure 7-11 on Page 7-14 to determine the weight and moment index of 370 gallons of Jet A fuel.

The CG under these loading conditions is located at station 296.2.

Determining the CG Shift When Cargo is Moved From One Section to Another

When cargo is shifted from one section to another, use this formula:

$$\Delta CG = \frac{\text{Weight shifted} \times \text{Distance shifted}}{\text{Total weight}}$$

If the cargo is moved forward, the CG is subtracted from the original CG. If it is shifted aft, add the CG to the original.

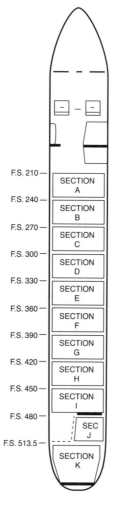

LOADING DATA
CARGO CONFIGURATION

SECTION	MAXIMUM STRUCTURAL CAPACITY	CENTROID ARM
A	600	F.S. 225
B	600	F.S. 255
C	600	F.S. 285
D	600	F.S. 315
E	600	F.S. 345
F	600	F.S. 375
G	600	F.S. 405
H	600	F.S. 435
I	600	F.S. 465
J	250	F.S. 499.5
K	565	F.S. 533

NOTES:

1. ALL CARGO IN SECTIONS A THROUGH J MUST BE SUPPORTED ON THE SEAT TRACKS AND SECURED TO THE SEAT TRACKS AND SIDE SEAT RAILS BY AN FAA-APPROVED SYSTEM.

2. CONCENTRATED CARGO LOADS IN SECTIONS A THROUGH L MUST NOT EXCEED 100 LBS PER SQUARE FOOT.

3. CARGO IN SECTIONS K AND L MUST BE RETAINED BY BAGGAGE WEBS AND PARTITIONS PROVIDED AS PART OF STANDARD AIRPLANE.

4. ANY EXCEPTIONS TO THE ABOVE PROCEDURES WILL REQUIRE APPROVAL BY A LOCAL FAA OFFICE.

Figure 7-14. *Loading data for cargo configuration.*

Item	Weight	Arm	Moment/100	CG
BOW	9,005		25,934	
Cargo Section A	300	225	675	
Cargo Section B	400	255	1,020	
Cargo Section C	450	285	1,283	
Cargo Section D	600	315	1,890	
Cargo Section E	600	345	2,070	
Cargo Section F	600	375	2,250	
Cargo Section G	200	405	810	
Cargo Section H		435		
Cargo Section I		465		
Cargo Section J		499.5		
Cargo Section K		533		
Fuel Jet A @ +15°C				
Gallons 370	2,516		7,520	
	14,671		43,452	296.2

Figure 7-15. *Flight manifest of a Beech 1900 in the cargo configuration.*

Determining the CG Shift When Cargo is Added or Removed

When cargo is added or removed, add or subtract the weight and moment index of the affected cargo to the original loading chart. Determine the new CG by dividing the new moment index by the new total weight, and multiply this by the reduction factor.

$$CG = \frac{\text{Total moment index}}{\text{Total weight}} \times \text{Reduction factor}$$

Determining Which Limits Are Exceeded

When preparing an aircraft for flight, you must consider all parameters and check to determine that no limit has been exceeded.

Consider the parameters below, and determine which limit, if any, has been exceeded.

- The airplane in this example has a basic empty weight of 9,005 pounds and a moment index of 25,934 pound-inches/100.

- The crew weight is 340 pounds, and its moment/100 is 439.

- The passengers and baggage have a weight of 3,950 pounds and a moment/100 of 13,221.

- The fuel is computed at 6.8 lbs/gal:

    The ramp load is 340 gallons, or 2,312 pounds.

    Fuel used for start and taxi is 20 gallons, or 136 pounds.

    Fuel remaining at landing is 100 gallons, or 680 pounds.

- Maximum takeoff weight is 16,600 pounds.

- Maximum zero fuel weight is 14,000 pounds.

- Maximum landing weight is 16,000 pounds.

Take these steps to determine which limit, if any, is exceeded:

1. Determine the zero fuel weight, which is the weight of the aircraft with all of the useful load except the fuel onboard.

Item	Weight	Moment/100	CG
Basic empty weight	9,005	25,934	
Crew	340	439	
PAX & Baggage	3,950	13,221	
Zero fuel weight	13,295	39,594	

The zero fuel weight of 13,295 pounds is less than the maximum of 14,000 pounds, so this parameter is acceptable.

2. Determine the takeoff weight and CG. The takeoff weight is the zero fuel weight plus the ramp load of fuel, less the fuel used for start and taxi. The takeoff CG is the (moment/100 ÷ weight) x 100.

Item	Weight	Moment/100	CG
Zero fuel weight	13,295	39,594	
Takeoff fuel 320 gal			
Ramp load – fuel			
for start & taxi			
340 – 20 = 320 gal	2,176	6,512	
Takeoff weight	15,471	46,106	298.0

The takeoff weight of 15,471 pounds is below the maximum takeoff weight of 16,600 pounds, and a check of Figure 7-12 on Page 7-14 shows that the CG at station 298.0 is also within limits.

3. Determine the landing weight and CG. This is the zero fuel weight plus the weight of fuel at landing.

Item	Weight	Moment/100	CG
Zero fuel weight	13,295	39,594	
Fuel at landing 100 gal	680	1,977	
Landing weight	13,975	41,571	297.5

The landing weight of 13,975 pounds is less than the maximum landing weight, of 14,000 to 16,000 pounds. According to Figure 7-12, the landing CG at station 297.5 is also within limits.

Use of Computer for
Weight and Balance Computations

Almost all weight and balance problems involve only simple math. This allows slide rules and hand-held electronic calculators to relieve us of much of the tedium involved with these problems. This chapter gives a comparison of the methods of determining the CG of an airplane while it is being weighed. First, determine the CG using a simple electronic calculator, then solve the same problem using an E6-B flight computer. Then, finally, solve it using a dedicated electronic flight computer.

Later in this chapter are examples of typical weight and balance problems (solved with an electronic calculator) of the kind that pilots and the A&P mechanics and repairmen will encounter throughout his or her aviation endeavors.

Using an Electronic Calculator to Solve Weight and Balance Problems

Determining the CG of an airplane in inches for the main-wheel weighing points can be done with any simple electronic calculator that has addition (+), subtraction (-), multiplication (x), and division (÷) functions. Scientific calculators with such additional functions as memory (M), parentheses (( )), plus or minus (+/-), exponential (y^x), reciprocal (1/x), and percentage (%) functions allow you to solve more complex problems or to solve simple problems using fewer steps.

The chart in Figure 8-1 includes data on the airplane used in this example problem.

Weighing Point	Weight (lbs)	Arm (in)
Right side	830	0
Left side	836	0
Nose	340	–78
Total	2,006	

Figure 8-1. *Weight and balance data of a typical nose wheel airplane.*

According to Figure 8-1, the weight of the nose wheel (F) is 340 pounds, the distance between main wheels and nose wheel (L) is -78 inches, and the total weight (W) of the airplane is 2,006 pounds. (L is negative because the nose wheel is ahead of the main wheels.)

To determine the CG, use this formula:

$$CG = \frac{F \times L}{W}$$
$$= \frac{340 \times -78}{2,006}$$

Key the data into the calculator as shown in red, and when the equal (=) key is pressed, the answer (shown here in green will appear.

$$(340)(x)(78)(+/-)(÷)(2006)(=) \ -13.2$$

The arm of the nose wheel is negative, so the CG is -13.2, or 13.2 inches ahead of the main-wheel weighing points.

Using an E6-B Flight Computer to Solve Weight and Balance Problems

The E6-B uses a special kind of slide rule. Instead of its scales going from 1 to 10, as on a normal slide rule, both scales go from 10 to 100. The E6-B cannot be used for addition or subtraction, but it is useful for making calculations involving multiplication and division. Its accuracy is limited, but it is sufficiently accurate for most weight and balance problems.

The same problem that was just solved with the electronic calculator can be solved on an E6-B by following these steps:

$$CG = \frac{F \times L}{W}$$
$$= \frac{340 \times -78}{2,006}$$

First, multiply 340 by 78 (disregard the - sign): [Figure 8-2a.]

- Place 10 on the inner scale (this is the index) opposite 34 on the outer scale (this represents 340) (Step 1).

- Opposite 78 on the inner scale, read 26.5 on the outer scale (Step 2).

- Determine the value of these digits by estimating: 300 x 80 = 24,000, so 340 x 78 =26,500.

  Then, divide 26,500 by 2,006: [Figure 8-2b.]

- On the inner scale, place 20, which represents 2,006, opposite 26.5 on the outer scale. (26.5 represents 26,500) (Step 3)

- Opposite the index, 10, on the inner scale read 13.2 on the outer scale (Step 4).

- Determine the value of 13.2 by estimating: 20,000 ÷ 2000 = 10, so 26,500 ÷ 2,006 = 13.2.

- The arm (-78) is negative, so the CG is also negative.

The CG is -13.2 inches, or 13.2 inches ahead of the datum.

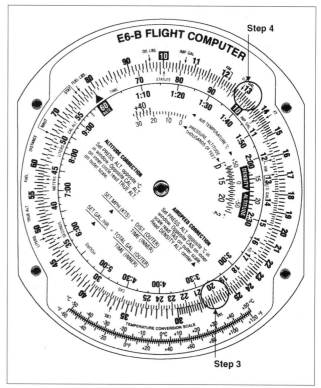

Figure 8-2b. *E6-B computer set up to divide 26,500 by 2,006.*

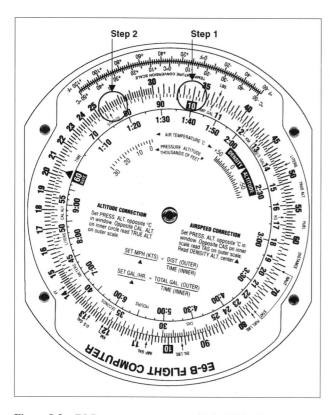

Figure 8-2a. *E6-B computer set up to multiply 340 by 78.*

Using a Dedicated Electronic Flight Computer to Solve Weight and Balance Problems

Dedicated electronic flight computers like the one in Figure 8-3 are programmed to solve many flight problems, such as wind correction, heading and ground speed, endurance, and true airspeed (TAS), as well as weight and balance problems.

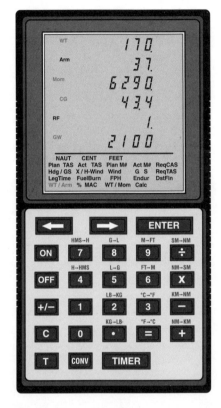

Figure 8-3. *Dedicated electronic flight computers are programmed to solve weight and balance problems, as well as flight problems.*

The problem just solved with an electronic calculator and an E6-B can also be solved with a dedicated flight computer using the information shown in Figure 8-1.

Each flight computer handles the problems in slightly different ways, but all are programmed with prompts that solicit you to input the required data so you do not need to memorize any formulas. Weight and arms are input as called for, and a running total of the weight, moment, and CG are displayed.

Typical Weight and Balance Problems

A hand-held electronic calculator like the one in Figure 8-4 is a valuable tool for solving weight and balance problems. It can be used for a variety of problems and has a high degree of accuracy. The examples given here are

solved with a calculator using only the (X),(÷),(+),(-), and (+/-) functions. If other functions are available on your calculator, some of the steps may be simplified.

Figure 8-4. *A typical electronic calculator is useful for solving most types of weight and balance problems.*

Determining CG in Inches From the Datum

This type of problem is solved by first determining the location of the CG in inches from the main-wheel weighing points, then measuring this location in inches from the datum. There are four types of problems involving the location of the CG relative to the datum.

Nose Wheel Airplane with Datum Ahead of the Main Wheels

The datum (D) is 128 inches ahead of the main-wheel weighing points, the weight of nose wheel (F) is 340 pounds, and the distance between main wheels and nose wheel (L) is 78 inches. The total weight (W) of the airplane is 2,006 pounds. Refer to Figure 3-5 on page 3-5.

Use this formula:

$$CG = D - \left(\frac{F \times L}{W}\right)$$

1. Determine the CG in inches from the main wheel:

   (340)(x)(78)(÷)(2006)(=) 13.2

2. Determine the CG in inches form the datum:

   (128)(-)(13.2)(=) 114.8

   The CG is 114.8 inches behind the datum.

Nose Wheel Airplane with Datum Behind the Main Wheels

The datum (D) is 75 inches behind the main-wheel weighing points, the weight of the nose wheel (F) is 340 pounds, and the distance between main wheels and nose wheel (L) is 78 inches. The total weight (W) of the airplane is 2,006 pounds. Refer to Figure 3-6 on page 3-5. Use this formula:

$$CG = -\left(D + \frac{F \times L}{W}\right)$$

1. Determine the CG in inches from the main wheels:

    (340)(x)(78)(÷)(2006)(=) 13.2

2. Determine the CG in inches from the datum:

    (75)(+)(13.2)(=) 88.2

    The minus sign before the parenthesis in the formula means the answer is negative. The CG is 88.2 inches ahead of the datum (-88.2).

Tail Wheel Airplane with Datum Ahead of the Main Wheels

The datum (D) is 7.5 inches ahead of the main-wheel weighing points, the weight of the tail wheel (R) is 67 pounds, and the distance between main wheels and tail wheel (L) is 222 inches. The total weight (W) of the airplane is 1,218 pounds. Refer to Figure 3-7 on page 3-6.

Use this formula:

$$CG = D + \left(\frac{R \times L}{W}\right)$$

1. Determine the CG in inches from the main wheels.

    (67)(x)(222)(÷)(1218)(=) 12.2

2. Determine the CG in inches from the datum:

    (7.5)(+)(12.2)(=) 19.7

    The CG is 19.7 inches behind the datum.

Tail Wheel Airplane with Datum Behind the Main Wheels.

The datum (D) is 80 inches behind the main-wheel weighing points, the weight of the tail wheel (R) is 67 pounds, and the distance between main wheels and tail wheel (L) is 222 inches. The total weight (W) of the airplane is 1,218 pounds. Refer to Figure 3-8 on page 3-6.

Use this formula:

$$CG = -D + \left(\frac{R \times L}{W}\right)$$

1. Determine the CG in inches from the main wheels:

    (67)(x)(222)(÷)(1218)(=) 12.2

2. Determine the CG in inches from the datum:

    (80)(+/-)(+)(12.2)(=) -67.8

    The CG is 67.8 inches ahead of the datum.

Determining CG, Given Weights and Arms

Some weight and balance problems involve weights and arms to determine the moments. Divide the total moment by the total weight to determine the CG. Figure 8-5 contains the specifications for determining the CG using weights and arms.

Determine the CG by using the data in Figure 8-5 and following these steps:

1. Determine the total weight and record this number:

    (830)(+)(836)(+)(340)(=)(2006)

2. Determine the moment of each weighing point and record them:

    (830)(x)(128)(=)106240
    (836)(x)(128)(=) 107008
    (340)(x)(50)(=) 17000

Weighing Point	Weight (lbs)	Arm (in)	Moment (lb-in)	CG
Right side	830	128	106,240	
Left side	836	128	107,008	
Nose	340	50	17,000	
Total	2,006		230,248	114.8

Figure 8-5. *Specifications for determining the CG of an airplane using weight and arm.*

3. Determine the total moment and divide this by the total weight:

$$(106240)(+)(107008)(+)(17000)(=)(÷)(2006)(=) \; 114.8$$

This airplane weighs 2,006 pounds and its CG is 114.8 inches from the datum.

Determining CG, Given Weights and Moment Indexes

Other weight and balance problems involve weights and moment indexes, such as moment/100, or moment/1,000. To determine the CG, add all the weights and all the moment indexes. Then divide the total moment index by the total weight and multiply the answer by the reduction factor. Figure 8-6 contains the specifications for determining the CG using weights and moments indexes.

Determine the CG by using the data in Figure 8-6 and following these steps:

1. Determine the total weight and record this number:

$$(830)(+)(836)(+)(340)(=) \; 2006$$

2. Determine the total moment index, divide this by the total weight, and multiply it by the reduction factor of 100:

$$(1,062.4)(+)(1,070.1)(+)(170)(=)(2302.5)(÷)(2006)(=)$$
$$(1.148)(x)(100)(=)114.8$$

This airplane weighs 2,006 pounds and its CG is 114.8 inches form the datum.

Weighing Point	Weight (lbs)	Moment/100	CG
Right side	830	1,062.4	
Left side	836	1,070.1	
Nose	340	170	
Total	2,006	2,302.5	114.8

Figure 8-6. *Specifications for determining the CG of an airplane using weights and moment indexes.*

Determining CG in Percent of Mean Aerodynamic Chord

- The loaded CG is 42.47 inches aft of the datum.
- MAC is 61.6 inches long.
- LEMAC is at station 20.1

1. Determine the distance between the CG and LEMAC:

$$(42.47)(-)(20.1)(=) \; 22.37$$

2. Then, use this formula:

$$CG \text{ in } \% \text{ MAC} = \frac{\text{Distance aft of LEMAC} \times 100}{\text{MAC}}$$

$$(22.37)(x) \, (100)(÷)(61.6)(=) \; 36.3$$

The CG of this airplane is located at 36.3% of the mean aerodynamic chord.

Determining Lateral CG of a Helicopter

It is often necessary when working weight and balance of a helicopter to determine not only the longitudinal CG, but the lateral CG as well. Lateral CG is measured from butt line zero (BL 0). All items and moments to the left of BL 0 are negative, and all those to the right of BL 0 are positive. Figure 8-7 contains the specifications for determining the lateral CG of a typical helicopter.

Determine the lateral CG by using the data in Figure 8-7 and following these steps:

1. Add all of the weights:

$$(1545)(+)(170)(+)(200)(+)(288)(=) \; 2203$$

2. Multiply the lateral arm (the distance between butt line zero and the CG of each item) by its weight to get the lateral offset moment of each item. Moments to the right of BL 0 are positive and those to the left are negative.

$$(1545)(x)(.2)(=) \; 309$$
$$(170)(x)(13.5)(+/-)(=) \; -2295$$
$$(200)(x)(13.5)(=) \; 2700$$
$$(288)(x)(8.4)(+/-)(=) \; -2419$$

Item	Weight	Lateral Arm	Lateral Offset Moment	Lateral CG
Helicopter empty weight	1,545	+0.2	309	
Pilot	170	−13.5	−2,295	
Passenger	200	+13.5	2,700	
Fuel 48 gal	288	−8.4	−2,419	
Total	2,203		−1,705	−0.77

Figure 8-7. *Specifications for determining the lateral CG of a helicopter.*

3. Determine the algebraic sum of the lateral offset moments.

$$(309)(+)(2295)(+/-)(+)(2700)(+)(2419)(+/-)(=) -1705$$

4. Divide the sum of the moments by the total weight to determine the lateral CG.

$$(1705)(+/-)(\div)(2203)(=) -0.77$$

The lateral CG is 0.77 inch to the left of butt line zero.

Determining ΔCG Caused by Shifting Weights

Fifty pounds of baggage is shifted from the aft baggage compartment at station 246 to the forward compartment at station 118. The total airplane weight is 4,709 pounds. How much does the CG shift?

1. Determine the number of inches the baggage is shifted:

$$(246)(-)(118)(=) 128$$

2. Use this formula:

$$\Delta CG = \frac{\text{Weight shifted} \times \text{Distance weight is shifted}}{\text{Total weight}}$$

$$(50)(x)(128)(\div)(4709)(=) 1.36$$

The CG is shifted forward 1.36 inches.

Determining Weight Shifted to Cause Specified ΔCG

How much weight must be shifted from the aft baggage compartment at station 246 to the forward compartment at station 118, to move the CG forward 2 inches? The total weight of the airplane is 4,709 pounds.

1. Determine the number of inches the baggage is shifted:

$$\text{Weight shifted} = \frac{\Delta CG \times \text{Total weight}}{\text{Distance weight is shifted}}$$

$$(246)(-)(118)(=) 128$$

2. Use this formula:

$$(2)(x)(4709)(\div)(128)(=) 73.6$$

Moving 73.6 pounds of baggage from the aft compartment to forward compartment will shift the CG forward 2 inches.

Determining Distance Weight is Shifted to Move CG a Specific Distance

How many inches aft will a 56-pound battery have to be moved to shift the CG aft by 1.5 inches? The total weight of the airplane is 4,026 pounds.

Use this formula:

$$\text{Distance weight is shifted} = \frac{\Delta CG \times \text{Total weight}}{\text{Weight shifted}}$$

$$(1.5)(x)(4026)(\div)(56)(=) 107.8$$

Moving the battery aft by 107.8 inches will shift the CG aft 1.5 inches.

Determining Total Weight of an Aircraft That Will Have a Specified ΔCG When Cargo is Moved

What is the total weight of an airplane if moving 500 pounds of cargo 96 inches forward shifts the CG 2.0 inches?

Use this formula:

$$\text{Total weight} = \frac{\text{Weight shifted} \times \text{Dist. weight is shifted}}{\Delta CG}$$

$$(500)(x)(96)(\div)(2)(=) 24000$$

Moving 500 pounds of cargo 96 inches forward will cause a 2.0-inch shift in CG of a 24,000-pound airplane.

Determining Amount of Ballast Needed to Move CG to a Desired Location

How much ballast must be mounted at station 228 to move the CG to its forward limit of +33? The airplane weighs 1,876 pounds and the CG is at +32.2, a distance of 0.8 inch out of limit.

Use this formula:

$$\text{Ballast weight} = \frac{\text{Aircraft empty weight} \times \text{Dist. out of limits}}{\text{Distance ballast to desired CG}}$$

$$(1876)(x)(.8)(\div)(195)(=) 7.7$$

Attaching 7.7 pounds of ballast to the bulkhead at station 228 will move the CG to +33.0.

Appendix

Supplemental Study Materials for Aircraft Weight and Balance

Advisory Circulars (check for most current revision)

AC 43.13-1B, Acceptable Methods, Techniques, and
    Practices - Aircraft Inspection and Repair

AC 65-9A, Airframe and Powerplant Mechanics General
    Handbook

AC 90-89A, Amateur-Built Aircraft and Ultralight Flight
    Testing Handbook

Other Study Materials

Aviation Maintenance Technician Series - General (AMT -G)
Aviation Supplies & Academics (ASA), Inc

Aircraft Basic Science (ABS)
*Glencoe Division, Macmillan/McGraw-Hill Publications
Company*

A & P Technician General Textbook (EA-ITP-G2)
Jeppesen-Sanderson, Inc.

Introduction to Aircraft Maintenance
Avotek Informantion Resources

Glossary

A note on glossary terms: over the years there has been a proliferation of aircraft weight and balance terms. This is the result of many factors such as: the FAA certification regulation an aircraft was constructed under, the FAA regulation the aircraft is being operated under, manufacturers standardization agreements, or a combination of these and others (an example are terms such as: Empty Weight, Licensed Empty Weight, Basic Empty Weight, Operational Empty Weight, and so on).

Many of these terms may have similar meanings or sound similar. Pilots and aircraft mechanics must ensure they understand the terminology and are applying the correct values based on the procedure and situation dictating the calculations undertaken.

In the Glossary, occasionally terms or a term word will be followed by the word GAMA in parentheses, this indicates that it is part of the standardized format adopted by the General Aviation Manufacturers Association in 1976 know as GAMA Specification No.1. These aircraft in general are manufactured under 14 CFR part 23 and differ from aircraft manufactured under the earlier certification rule Civil Air Regulation Part 3 for weight and balance purposes in the condition under which empty weight was established.

The aircraft that are certified per 14 CFR parts 23, 25, 27, and 29 established their certificated empty weight as required in the appropriate section of these regulations which to paraphrase states: The empty weight and corresponding center of gravity must be determined by weighting the aircraft with:

- Fixed ballast
- Unusable fuel
- Full operating fluid, including, oil, hydraulic fluid, and other fluids required for normal operation of the aircraft systems, except potable water, lavatory precharge water, and water intended for injection in the engine(s).

In contrast aircraft certificated under the older Civil Air Regulations (CARs) established certificated empty weight under similar conditions as the newer aircraft with the important exception that the aircraft weight did not include full oil, only undrainable oil. Mechanics and repair stations should consult the appropriate certification rule when reestablishing empty weight.

14 CFR part 121. The Federal regulations governing domestic, flag, and supplemental operations.

14 CFR part 135. The Federal regulations governing Commuter and On-Demand Operations.

Adverse Loaded CG Check. A weight and balance check to determine that no condition of legal loading of an aircraft can move the CG outside of its allowable limits.

Aircraft Specifications. Documentation containing the pertinent specifications for aircraft certificated under the CARs.

Airplane Flight Manual (AFM). An FAA-approved document, prepared by the holder of a Type Certificate for an aircraft, that specifies the operating limitations and contains the required markings and placards and other information applicable to the regulations under which the aircraft was certificated.

Approved Type Certificate. A certificate of approval issued by the FAA for the design of an aircraft, engine, or propeller.

Arm. (GAMA) The horizontal distance from the reference datum to the center of gravity (CG) of an item. The algebraic sign is plus (+) if measured aft of the datum or to the right side of the center line when considering a lateral calculation. The algebraic sign is minus (-) if measured forward of the datum or the left side of the center line when considering a lateral calculation.

Balanced Laterally. Balanced in such a way that the wings tend to remain level.

Ballast. A weight installed or carried in an aircraft to move the center of gravity to a location within its allowable limits.

Permanent Ballast (fixed ballast). A weight permanently installed in an aircraft to bring its center of gravity into allowable limits. Permanent ballast is part of the aircraft empty weight.

Temporary Ballast. Weights that can be carried in a cargo compartment of an aircraft to move the location of CG for a specific flight condition. Temporary ballast must be removed when the aircraft is weighed.

Basic Empty Weight. (GAMA) Standard empty weight plus optional equipment.

Basic Operating Index. The moment of the airplane at its basic operating weight divided by the appropriate reduction factor.

Basic Operating Weight (BOW). The empty weight of the aircraft plus the weight of the required crew, their baggage and other standard item such as meals and potable water.

Bilge Area. The lowest part of an aircraft structure in which water and contaminants collect.

Butt (or buttock) Line Zero. A line through the symmetrical center of an aircraft from nose to tail. It serves as the datum for measuring the arms used to determine the lateral CG. Lateral moments that cause the aircraft to rotate clockwise are positive (+) , and those that cause it to rotate counterclockwise are negative (-).

Calendar Month. A time period used by the FAA for certification and currency purposes. A calendar month extends from a given day until midnight of the last day of that month.

Civil Air Regulation (CAR). predecessor to the Federal Aviation Regulations.

CAMs. The manuals containing the certification rules under the Civil Air Regulations.

Center of Gravity (CG). (GAMA) The point at which an airplane would balance if suspended. Its distance from the reference datum is determined by dividing the total moment by the total weight of the airplane. It is the mass center of the aircraft, or the theoretical point at which the entire weight of the aircraft is assumed to be concentrated. It may be expressed in percent of MAC (mean aerodynamic cord) or in inches from the reference datum.

Center of Lift. The location along the chord line of an airfoil at which all the lift forces produced by the airfoil are considered to be concentrated.

Centroid. The distance in inches aft of the datum of the center of a compartment or a fuel tank for weight and balance purposes.

CG Arm. (GAMA) The arm obtained by adding the airplane's individual moments and dividing the sum by the total weight.

CG Limits. (GAMA) The extreme center of gravity locations within which the aircraft must be operated at a given weight. These limits are indicated on pertinent FAA aircraft type certificate data sheets, specifications, or weight and balance records.

CG Limits Envelope. An enclosed area on a graph of the airplane loaded weight and the CG location. If lines drawn from the weight and CG cross within this envelope, the airplane is properly loaded.

CG Moment Envelope. An enclosed area on a graph of the airplane loaded weight and loaded moment. If lines drawn from the weight and loaded moment cross within this envelope, the airplane is properly loaded.

Chord. A straight-line distance across a wing from leading edge to trailing edge.

Curtailment. An operator created and FAA-approved operational loading envelope that is more restrictive than the manufacturer's CG envelope. It ensures that the aircraft will be operated within limits during all phases of flight. Curtailment typically accounts for, but is not limited to, in-flight movement of passengers and crew, service equipment, cargo variation, seating variation, ect.

Delta Δ. This symbol, Δ, means a change in something. ΔCG means a change in the center of gravity location.

Dynamic Load. The actual weight of the aircraft multiplied by the load factor, or the increase in weight caused by acceleration.

Empty Weight. The weight of the airframe, engines, all permanently installed equipment, and unusable fuel. Depending upon the part of the federal regulations under which the aircraft was certificated, either the undrainable oil or full reservoir of oil is included.

Empty-weight Center of Gravity (EWCG). This is the center of gravity of the aircraft in the empty condition, containing only the items specified in the aircraft empty weight. This CG is an essential part of the weight and balance record of the aircraft.

Empty-weight Center of Gravity Range. The distance between the allowable forward and aft empty-weight CG limits.

Equipment List. A list of items approved by the FAA for installation in a particular aircraft. The list includes the name, part number, weight, and arm of the component. Installation or removal of an item in the equipment list is considered to be a minor alteration.

Fleet Weight. An average weight accepted by the FAA for aircraft of identical make and model that have the same equipment installed. When a fleet weight control program is in effect, the fleet weight of the aircraft can be used rather than every individual aircraft having to be weighed.

Fuel Jettison System. A fuel subsystem that allows the flight crew to dump fuel in an emergency to lower the weight of an aircraft to the maximum landing weight if a return to landing is required before sufficient fuel is burned off. This system must allow enough fuel to be jettisoned that the aircraft can still meet the climb requirements specified in 14 CFR part 25.

Fulcrum. The point about which a lever balances.

Index Point. A location specified by the aircraft manufacturer from which arms used in weight and balance computations are measured. Arms measured from the index point are called index arms.

Interpolate. To determine a value in a series between two known values.

Landing Weight. The takeoff weight of an aircraft less the fuel burned and/or dumped en route.

Large Aircraft (14 CFR part 1). An aircraft of more than 12,500 pounds, maximum certificated takeoff weight.

Lateral Balance. Balance around the roll, or longitudinal, axis.

Lateral Offset Moment. The moment, in lb-in, of a force that tends to rotate a helicopter about its longitudinal axis. The lateral offset moment is the product of the weight of the object and its distance from butt line zero. Lateral offset moments that tend to rotate the aircraft clockwise are positive, and those that tend to rotate it counterclockwise are negative.

LEMAC. Leading Edge of the Mean Aerodynamic Chord.

Load Cell. A component in an electronic weighing system that is placed between the jack and the jack pad on the aircraft. The load cell contains strain gauges whose resistance changes with the weight on the cell.

Load Factor. The ration of the maximum load an aircraft can sustain to the total weight of the aircraft. Normal category aircraft must have a load factor of a least 3.8, Utility category aircraft 4.4, and acrobatic category aircraft, 6.0.

Loading Graph. A graph of load weight and load moment indexes. Diagonal lines for each item relate the weight to the moment index without having to use mathematics.

Loading Schedule. A method for calculating and documenting aircraft weight and balance prior to taxiing, to ensure the aircraft will remain within all required weight and balance limitations throughout the flight.

Weight and Balance Definitions

Single-engine aircraft:

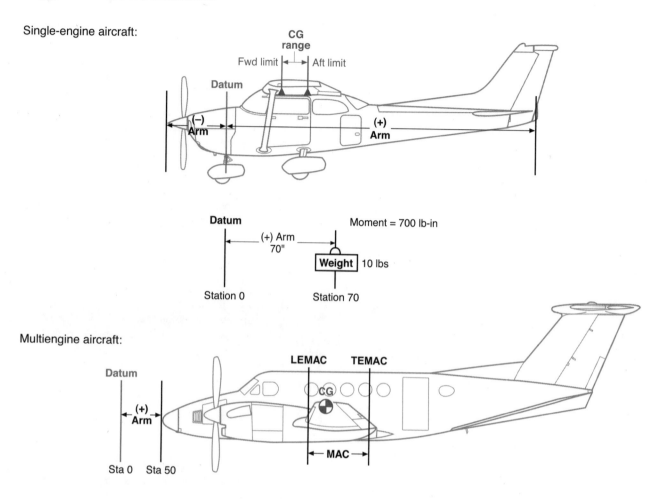

Multiengine aircraft:

Longitudinal Axis. An imaginary line through an aircraft from nose to tail, passing through its center of gravity.

Longitudinal Balance. Balance around the pitch, or lateral, axis.

MAC. Mean Aerodynamic Chord.

Major Alteration. An alteration not listed in the aircraft, aircraft engine, or propeller specifications, (1) that might appreciably affect weight, balance, structural strength, performance, powerplant operation, flight characteristics, or other qualities affecting airworthiness; or (2) that is not done according to accepted practices or cannot be done by elementary operations.

Maximum Landing Weight. (GAMA) Maximum weight approved for the landing touchdown.

Maximum Permissible Hoist Load. The maximum external load that is permitted for a helicopter to carry. This load is specified in the POH.

Maximum Ramp Weight. (GAMA) Maximum weight approved for ground maneuver. It includes weight of start, taxi, and runup fuel.

Maximum Takeoff Weight. (GAMA) Maximum weight approved for the start of the takeoff run.

Maximum Taxi Weight. Maximum weight approved for ground maneuvers. This is the same as maximum ramp weight.

Maximum Weight. The maximum authorized weight of the aircraft and all of its equipment as specified in the Type Certificate Data Sheets (TCDS) for the aircraft.

Maximum Zero Fuel Weight. The maximum authorized weight of an aircraft without fuel. This is the total weight for a particular flight less the fuel. It includes the aircraft and everything that will be carried on the flight except the weight of the fuel.

METO Horsepower (maximum except takeoff). The maximum power allowed to be continuously produced by an engine. Takeoff power is usually limited to a given amount of time, such as 1 minute or 5 minutes.

Minimum Fuel. The amount of fuel necessary for one-half hour of operation at the rated maximum-continuous power setting of the engine, which, for weight and balance purposes, is 1/12 gallon per maximum-except-takeoff (METO) horse-power. It is the maximum amount of fuel that could be used in weight and balance computations when low fuel might adversely affect the most critical balance conditions. To determine the weight of the minimum fuel in pounds, divide the METO horsepower by two.

Minor Alteration. An alteration other than a major alteration. This includes alterations that are listed in the aircraft, aircraft engine, or propeller specifications.

Moment. A force that causes or tries to cause an object to rotate. It is indicated by the product of the weight of an item multiplied by its arm.

Moment. (GAMA) The product of the weight of an item multiplied by its arm. (Moment divided by a constant is used to simplify balance calculations by reducing the number of digits; see reduction factor.)

Moment Index. The moment (weight times arm) divided by a reduction factor such as 100 or 1,000 to make the number smaller and reduce the chance of mathematical errors in computing the center of gravity.

Moment Limits vs. Weight Envelope. An enclosed area on a graph of three parameters. The diagonal line representing the moment/100 crosses the horizontal line representing the weight at the vertical line representing the CG location in inches aft of the datum. When the lines cross inside the envelope, the aircraft is loaded within its weight and CG limits.

Net Weight. The weight of the aircraft less the weight of any chocks or other devices used to hold the aircraft on the scales.

Normal Category. A category of aircraft certificated under 14 CFR part 23 and CAR part 3 that allows the maximum weight and CG range while restricting the maneuvers that are permitted.

PAX. Passengers.

Payload. (GAMA) Weight of occupants, cargo, and baggage.

Pilot's Operating Handbook (POH). An FAA-approved document published by the airframe manufacturer that lists the operating conditions for a particular model of aircraft and its engine(s).

Potable Water. Water carried in an aircraft for the purpose of drinking.

Ramp Weight. The zero fuel weight plus all of the usable fuel on board.

Reference Datum. (GAMA) An imaginary vertical plane from which all horizontal distances are measured for balance purpose.

Reduction Factor. A number, usually 100 or 1,000 by which a moment is divided to produce a smaller number that is less likely to cause mathematical errors when computing the center of gravity.

Residual Fuel. Fuel that remains trapped in the system after draining the fuel from the aircraft with the aircraft in level flight attitude. The weight of this residual fuel is counted as part of the empty weight of the aircraft.

Service Ceiling. The highest altitude at which an aircraft can still maintain a steady rate of climb of 100 feet per minute.

Small Aircraft (14 CFR part 1). An aircraft weighing 12,500 pounds or less, maximum certificated takeoff weight.

Standard Empty Weight. (GAMA) Weight of a standard airplane including unusable fuel, full operating fluids, and full oil.

Static Load. The load imposed on an aircraft structure due to the weight of the aircraft and its contents.

Station. (GAMA) A location along the airplane fuselage usually given in terms of distance from the reference datum.

Strain Sensor. A device that converts a physical phenomenon into an electrical signal. Strain sensors in a wheel axle sense the amount the axle deflects and create an electrical signal that is proportional to the force that caused the deflection.

Structural Station. This is a location in the aircraft, such as a bulkhead, which is identified by a number designating its distance in inches or percent MAC from the datum. The datum is, therefore, identified as station zero. The stations and arms are identical. An item located at station +50 would have an arm of 50 inches.

Takeoff Weight. The weight of an aircraft just before beginning the takeoff roll. It is the ramp weight less the weight of the fuel burned during start and taxi.

Tare Weight. The weight of any chocks or devices that are used to hold an aircraft on the scales when it is weighed. The tare weight must be subtracted from the scale reading to get the net weight of the aircraft.

TEMAC. Trailing Edge of the Mean Aerodynamic Chord.

Type Certificate Data Sheets (TCDS). The official specifications issued by the FAA for an aircraft, engine, or propeller.

Undrainable Oil. Oil that does not drain from an engine lubricating system when the aircraft is in the normal ground attitude and the drain valve is left open. The weight of the undrainable oil is part of the empty weight of the aircraft.

Unusable Fuel. (GAMA) Fuel remaining after a runout test has been completed in accordance with governmental regulations.

Usable Fuel. (GAMA) Fuel available for flight planning.

Useful Load. (GAMA) Difference between takeoff weight, or ramp weight if applicable, and basic empty weight.

Utility Category. A category of aircraft certificated under 14 CFR part 23 and CAR part 3 that permits limited acrobatic maneuvers but restricts the weight and the CG range.

Wing Chord. A straight-line distance across a wing from leading edge to trailing edge.

Zero Fuel Weight. The weight of an aircraft without fuel.

Index